Economics, Entropy and the Environment

T. Randolph Beard: To my mother and father
Gabriel A. Lozada: To my wonderful family

Economics, Entropy and the Environment:

The Extraordinary Economics of Nicholas Georgesçu-Roegen

T. Randolph Beard

Associate Professor of Economics, Auburn University, USA

Gabriel A. Lozada

Associate Professor of Economics, The University of Utah, USA

Edward Elgar

Cheltenham, UK • Northampton, MA, USA

Published by
Edward Elgar Publishing Limited
Glensanda House
Montpellier Parade
Cheltenham
Glos GL50 1UA
UK

Edward Elgar Publishing, Inc.
136 West Street
Suite 202
Northampton
Massachusetts 01060
USA

A catalogue record for this book is available from the British Library

Library of Congress Cataloguing in Publication Data

Beard, T. Randolph (Thomas Randolph)
 Economics, entropy and the environment: the extraordinary economics
 of Nicholas Georgescu-Roegen / T. Randolph Beard, Gabriel A. Lozada.
 Includes bibliographical references and index.
 1. Georgescu-Roegen, Nicholas. 2. Economists—United States.
 3. Environmental economics. 4. Entropy. 5. Human ecology. I. Lozada,
 Gabriel A., 1959–. II. Title.
 HB119.G43B4   1999
 333.7—dc21
                                                  99–39679
                                                  CIP

Printed and bound in Great Britain by Bookcraft (Bath) Ltd.

ISBN 1 84064 122 3

Contents

About the authors

T. Randolph Beard was born in Durham, North Carolina, and raised in Baton Rouge, Louisiana. He received an undergraduate honors degree in economics from Tulane University, and a PhD in economics from Vanderbilt University, where he knew Nicholas Georgesçu-Roegen. Since 1988 he has been on the faculty of Auburn University, where he teaches courses in microeconomics and industrial organization. He is an author of *Initial Public Offerings: Findings and Theories* (Kluwer, 1995). His work has been published in the *Review of Economics and Statistics*, the *RAND Journal of Economics*, the *Journal of Business*, *Management Science*, and many other outlets.

Gabriel A. Lozada was born in Evanston, Illinois, and raised in Baton Rouge, Louisiana. While an undergraduate student at Louisiana State University, he took two classes from Professor Herman E. Daly, who introduced him to the work of Nicholas Georgesçu-Roegen. Lozada earned a BS degree in physics *summa cum laude* from Louisiana State University in 1981, and a BA degree in economics *summa cum laude* from the same university in the same year. He attended graduate school at Stanford University, where he earned an MS degree in engineering–economic systems in 1983, an MA degree in economics in 1984, and a PhD degree in economics in 1987. He is currently an Associate Professor of Economics at the University of Utah. He previously taught at Texas A&M University, and was a Ciriacy-Wantrup Postdoctoral Fellow at the University of California at Berkeley in 1991–2. Lozada's research concentrates on the economic theory of exhaustible resource industries, and he has published papers in journals such as *Resource and Energy Economics*, the *Journal of Economic Dynamics and Control*, the *International Economic Review*, and the *American Economic Review*.

Acknowledgements

A book of this kind is impossible without the assistance of many altruistic souls. We wish to thank, without implicating in any way, Herman Daly, Craufurd Goodwin, Elton Hinshaw, V. Kerry Smith, Terry Anderson, Gary Becker, Roy Ruffin, Jim Leitzel, Kozo Mayumi, John Siegfried, Chris Klein, Bob Ekelund, Bob Hébert, Leland Yeager, Harris Schlesinger, Malte Faber, the professors of economics at the University of Utah, especially Allen Sievers. We should also like to thank Edward Elgar and Dymphna Evans at Edward Elgar Publishing.

Beard gratefully acknowledges manuscript assistance from Lola Burnett, Faith Pence, Holly Kliest-Constantino, Beth Rush, Deede Smith and Carla Hicks.

The cover artwork is an original watercolour of Georgesçu-Roegen by Bob Ekelund.

We particularly thank our wives, Jessie and Leslie, for putting up with us as we worked on this book.

Foreword

Georgesçu-Roegen was a truly great economist, one from whom the present generation of economists still has very much to learn. This welcome exposition of his major ideas by Beard and Lozada should help economists understand Georgesçu, both the revolutionary boldness and originality of many of his ideas and the careful logic with which he developed them. I believe Beard and Lozada have done an excellent job, both of selection and of exposition. This was no easy task since Georgesçu's writings range from the theory of the mathematical continuum to the sociology of the peasant village. My hope is that this book will do for Georgescu's ideas what Alvin Hansen did for the ideas of John Maynard Keynes with his *A Guide to Keynes*, published back in 1953. In my view Georgesçu's intellectual contribution will eventually prove to be even greater than that of Keynes. We have a long way to go before Georgesçu's ideas exert even a tenth of the influence that Keynes's ideas have had. But as any student of Georgesçu knows, fundamental qualitative change and emergent novelty rule the world.

I confess to a more personal pleasure at the publication of this book as well. One of the few delights of growing older is that of learning from one's former students. Gaby Lozada was an undergraduate student of mine at Louisiana State University (LSU), and was introduced to Georgesçu's writings in my course. I first explained the easy parts of Georgesçu to him, and now he explains the hard parts to me. I got the better of that deal. While Randy Beard was never my student I knew him slightly as a youngster because he is the son of my former LSU colleague, Tommy Beard. By coincidence Randy studied economics at Vanderbilt, where I had been Georgesçu's student many years before. Georgesçu was still around, although retired. To his credit Randy recognized genius when he saw it and maximized his contact with Georgesçu, even though I doubt that such contact was encouraged by the then active economics faculty. I am not sure just how these co-authors got together, but I am glad they did, and I am sure that many readers will share my delight and gratitude for the ripe fruits of their intellectual collaboration.

Herman E. Daly

1.   Introduction

In their fascinating text, *An Outline of the History of Economic Thought*, Ernesto Screpanti and Stefano Zamagni conclude their narrative with a small section entitled "Four unconventional economists." In introducing this section, the authors comment that "There are four economists for whom we have not found the right place in the panorama of contemporary economic schools . . . ."[1] These unconventional economists are John K. Galbraith, A.O. Hirschman, Richard Goodwin, and Nicholas Georgeşçu-Roegen. Screpanti and Zamagni note: "We have put them together because we are convinced that their resistance to classification is a characteristic that unites and defines them more clearly than might first appear. And we have put them with the heretics as we believe that, among the qualities that unite them, the taste of heresy is not the least important."[2]

It is a primary purpose of this book to try and find the "right place" for Nicholas Georgeşçu-Roegen. This is by no means an easy task, for reasons that will become apparent. Nevertheless, we are strongly of the opinion that Georgeşçu is an economist whose work is important, original, and largely correct. This view of Georgeşçu is itself somewhat heretical. It is common for conventionally trained economists, especially those under the age of 50, to give relatively little credence to Georgeşçu. Many younger students of economics are totally unfamiliar with Georgeşçu's "orthodox" contributions to utility and production theory and may only vaguely recall the various controversies surrounding the 1971 publication of *The Entropy Law and the Economic Process*.[3]

Georgeşçu-Roegen's application of thermodynamics to economics, combined with his often impenetrable writing style and his sometimes abrasive approach to colleagues, have resulted in a minimization of Georgeşçu-Roegen's impact in modern economics. We believe that this result is both unfortunate and unjustified, and this book constitutes our argument.

The historian of economic thought, Professor Philip Mirowski, dedicated his book, *More Heat Than Light*, to Nicholas Georgeşçu-Roegen. This was an appropriate tribute because Georgeşçu, perhaps more than any other modern economist, thought and wrote deeply on matters epistemological and methodological throughout much of his career. Georgeşçu-Roegen sought always after the "true facts" (as opposed to the stylized ones) of economic life, and tried to identify all of those factors required to construct a valid representation of the economic process. He combined, in one mind, considerable technical and mathematical sophistication with the commonsensical, unadorned outlook of the practical man. Many of Georgesçu's most important insights arose from his unwillingness to accommodate sloppy, metaphorical reasoning. "What does this variable represent?", he would ask, adding, "In what units is it measured?" Such questions can be far more embarrassing than one ordinarily would like to admit.

Although many readers may associate Georgesçu-Roegen primarily with his invocation of the Second Law of Thermodynamics (the "Entropy Law") in economic analysis, it is an important goal of this book to explain that this association was the inevitable result of the methodological outlook Georgesçu held throughout his career. It is incorrect to say that the Entropy Law led Georgesçu to his unique logical system, termed "dialectics." Rather, dialectics led Georgesçu to the Entropy Law, and one may easily find references to dialectical reasoning, "evolution", and the inadequacy of various "mechanical" models of human behavior in his work predating the 1960s.

While it is fashionable now to use the word "entropy" in a variety of ways and to associate the term with fields such as information, chaos, and systems theory, entropy to Georgesçu referred strictly to an index of unavailable energy in a thermodynamic setting. Georgesçu-Roegen disparaged the use of the term "entropy" in information theory and other areas having nothing to do with temperature.[4]

Georgesçu's fascination with, and emphasis on, the Entropy Law arose at least partly from the unique forms this law takes. Common statements of the law include (a) heat always flows by itself from the warmer to the colder body; and (b) the entropy of an isolated system never decreases through time. As the forms of these statements illustrate, the Second Law of Thermodynamics is utterly unlike other physical laws. In fact, Georgesçu calls such laws as this "evolutionary," and argues with great force that such laws are valid, relevant, and illustrative of the inadequacy of mechanical (analytic) representations of reality. Georgesçu felt that, were economists more familiar with thermodynamics, they would be less the prisoners of their mechanical analogies. And the mechanistic prison in which economics finds itself renders economists incapable of fulfilling their most important roles as facilitators of a discussion of the best aims for mankind.[5]

In contemplating the Entropy Law, Georgesçu-Roegen found a deep and convincing confirmation of the importance of several ideas, or "themes", that had steadily infiltrated his work and his thinking over many years. These themes include the limitations of "binary" logic, the importance of qualitative change in all but the simplest systems, and the inadequacy of real numbers for modeling economic and almost all other complex phenomena. A good example of these ideas occurs in his famous article "The Nature of Expectation and Uncertainty"

(1958), in which he seriously evaluates the feasibility of adequately representing expectations with numbers. A related example, described in the same work, is easier to recount.

Suppose a random variable X can take on any real value, $X \in R$, and let $f(X)$ be the marginal density function (measure) of X on R. Let X^0 be a particular value of X. It is quite common for students of statistics to be confused by the idea that the statements "$\text{Prob}(X = X^0) = 0$" and "$X = X^0$ cannot occur" are not equivalent and, indeed, the first may be true and the second false. This "problem" is endemic to statistics itself, and it arises from our requirement that the number zero do "double duty," representing both a zero probability and a logical impossibility.

Although these considerations may appear far removed from thermodynamics, at a very deep level Georgesçu-Roegen felt that they were closely related. Georgesçu used the term "arithmomorphic" to describe models that are constructed of functions of real numbers. Time enters such models, if at all, as "clock time," or elapsed mechanical time since some arbitrary date. All such models are incapable of representing "real" change, that is, evolution, because the real numbers, and analytic relations among them, are capable only of representing *mechanical* analogies. For example, Newton's laws of motion and force and the relationships derived from them, are "time symmetric": if one showed a film of colliding billiard balls (on a frictionless surface) in reverse, nothing seen by the observer would indicate the error made in loading the film.

The Entropy Law, however, is of an entirely different character. This Law is *not* time symmetric (entropy increases as time flows, never the opposite) and, importantly, this law is not an analytic formula. The Second Law does not state how fast entropy must increase, only that it must. In his popular science monograph, *Five Equations that Changed the World*, Michael Guillen (1995) lists Newton's Law of Gravity, Bernoulli's Law of Hydrostatic Pressure, Faraday's Law of Electromagnetic Induction, Einstein's $E = mc^2$, and the Second Law of Thermodynamics. Yet, the Entropy Law is *not* an equation. An equation is a conservation rule or equivalence relation: the Second Law is an inequality which does not state what specifically must happen, but rather prohibits many possible paths. These prohibited future paths are not merely "zero probability" or "low probability" outcomes (such as $X = X^0$), but are actually *impossible*, according to Georgesçu-Roegen.

Besides believing that the Entropy Law in its epistemological sense was important to economic theory, Georgesçu-Roegen believed that the Entropy Law, in its phenomenological sense, was relevant to economic activity. Of course, no one can deny that economic activity involves physical activity and, as such, must be subject to all the laws of elementary matter. Economic activities cannot violate the laws of physics. Yet, Georgesçu's insight is correct in a more important, and more immediate, sense than is first apparent. First, the Second

Law implies that energy may not be recycled. Second, although it is not often stated in this fashion, the Second Law of Thermodynamics imposes a minimum energy requirement on the critical economic act of changing the naturally occurring, stable state of any material. This requirement is stated as an inequality constraint which *must* be strict. No technological advances could ever, even in principle, relieve us of this energy requirement, although advances could provide wondrous new ways of meeting this requirement. Further, having achieved our transformation of a material into a non-stable equilibrium state, the Second Law ordains an unending struggle to maintain the material in that state, as anyone who has owned a rusted automobile knows all too well.

Georgesçu-Roegen also argued, although somewhat inconsistently, that the Entropy Law applies to organized structures. Of course, the concept of entropy itself applies to matter, not energy, because only materials have an entropy.[6] However, this is not all that Georgesçu had in mind. Rather, he sometimes argued that "matter arranged in some definite structures" underwent irrevocable dissipation. One source of this belief was apparently the role of friction in any mechanical system. Experience suggests that, no matter how hard we try, all mechanical devices are subject to friction and eventually fail to function.

Georgesçu-Roegen's belief that the Entropy Law applied to organized structures is not tenable as stated, however. Georgesçu himself came to recognize this point, and instead proposed a new thermodynamic law, the "Fourth Law." This claim, which is unresolved but may be false, holds that no closed system that receives energy (but not matter) at a constant rate from outside can perform work at a constant rate forever. This law, if true, would imply that perfect recycling was in principle impossible as well. Biologists tend to doubt the validity of this law, as does Georgesçu's student Kozo Mayumi.[7] We are unsure of its validity.

Regardless of the judgment science may ultimately render on the Fourth Law, Georgesçu clearly has a point about material dissipation. One may go farther and summarize Georgesçu's economic insight quite simply as follows. The economic process is one in which stocks of some things are depleted, while stocks of other things accumulate. Things depleted include propitiously situated, useful material substances, while things accumulated include toxic by-products of industrial activity. Except for trivial quantities of metals, say, lost due to spacecraft launches, the actual amount of copper on Earth is today the same as it was 1 000 000 years ago. What has changed in the interim is how, and in what forms, that copper is distributed in the environment.

Considerations of entropy, material dissipation, and the like led Georgesçu-Roegen to formulate a new field of economics, dubbed "bioeconomics." Bioeconomics serves as an early and very interesting example of the notable modern trend of combining different scientific disciplines in order to analyze a system that involves great complexity. Bioeconomics combines economics, thermodynamics, biology, anthropology, sociology and political science in a whole intended to provide a basis for useful discussions of the "best aims and means for Mankind." Georgesçu-Roegen even offered a few models of a bioeconomic character late in his career, although he never completed an advertised book on this subject, and his papers appear to contain nothing that one could identify as a part of such a manuscript.[8]

As is clear from our discussion so far, Georgesçu-Roegen walked where most economists fear to tread. The application of thermodynamics to economic analysis and the proposal of a new thermodynamic law are intellectually courageous, to say the least. One might go further and dismiss Georgesçu-Roegen as an eccentric unworthy of serious study. This response would be a serious mistake. First, unlike many other notable economic "heretics," Nicholas Georgesçu-Roegen had very substantial neoclassical credentials stretching back to the 1930s. Georgesçu's contributions to utility theory and production models were fundamental. Georgesçu was consistently both rigorous and highly original, and he turned his attention to most of the important problems of his times. No other heretic of twentieth-century economics achieved quite the same level of recognition and awards from the establishment: fellowships in the American Economic Association and Econometric Society, the Richard Ely Lectureship, the Earl Sutherland Prize, an editorship of *Econometrica*, nominations for the Nobel prize, and so on.

Additionally, and more fundamentally, we believe that Georgesçu-Roegen should be taken seriously by economists, and should be studied carefully, because, by and large, he was correct in his views. Indeed, he was often right about problems when almost everyone else was wrong. For example, Georgesçu-Roegen, though employed as an economist and trained as a mathematician, rejected on technical grounds Boltzman's so-called "H-Theorem" in statistical thermodynamics at a time when many scientists accepted its validity. History has proven Georgesçu correct. Georgesçu's lack of physics training caught up with him, however, as his rejection of the "H-Theorem" led him further to reject the whole of statistical thermodynamics, an unfortunate error. Georgesçu was not infallible, but he was brilliant, original, and dedicated to the proposition that, as Joan Robinson put it, economics is a "serious subject."

This book is intended to introduce and explain the extraordinary work of Nicholas Georgesçu-Roegen to conventionally trained economists and others interested in environmental economics, ecology, or the history of science. Our treatment will be sympathetic but not sentimental. There is quite enough that is right and revolutionary in Georgesçu's work that we feel no compulsion to gild his achievements. When we believe Georgesçu is in error, we will say so. Our selection of material for coverage, while aiming for completeness, admittedly reflects our own views to a certain extent. The reader is now and always urged to read Georgesçu-Roegen's writings directly and to form an independent judgment. If we stimulate a discussion of Georgesçu's work, we will have succeeded in our work.

The book has the following outline. Chapter 2 presents a brief biographical sketch of Georgesçu-Roegen. Although Georgesçu had a very interesting, sometimes unpleasant and occasionally dangerous life, our purpose is less to entertain than to provide the background necessary for understanding Georgesçu's intellectual development. Georgesçu-Roegen was an economist of the 1930s–1950s, intellectually speaking, and his interests closely paralleled many of the important issues of those times. As a consequence, both his choice of research topics and his terminology are often confusing to modern readers. The fact that he didn't know English until he was in his late twenties probably does not help.

Chapter 3 reviews Georgesçu's economic methodology. Georgesçu-Roegen's greatest contributions to economics may well turn out to be methodological. He was a very harsh critic of the over-mathematicalization of economics, although the bases of his criticisms were mathematical. In particular, the widespread use of mechanical analogies, termed "arithmomorphism" by Georgesçu, has caused economists to virtually ignore entire classes of economic phenomena of an evolutionary character.

Chapter 4 offers a selected review and evaluation of Georgesçu-Roegen's contributions to what Kuhn termed "Normal Science", that is, research within the prevailing paradigm. Georgesçu made numerous important contributions in many areas. Most impressively, he solved the "integrability problem" in utility theory. Integrability refers to the problem of finding a preference field that generates a given demand system, and identifying those mathematical conditions that are necessary for the task. Georgesçu did so but, in the process, he broke wholly new ground in the areas of stochastic choice models and economic methodology. Postulate A of his famous paper, "The Pure Theory of Consumer's Choice" (1936), ruled out the "ordinalist's fallacy" (the incorrect belief that any set that can be completely ordered can be placed in one-to-one correspondence with the real number line). This idea became extremely important in Georgesçu's later work.

Georgesçu made many other contributions to utility theory, and additionally found important results, including the so-called "Substitution Theorem", in Leontief-type production models. Georgesçu's greatest contribution to production theory, however, is his flow-fund model. This innovation is described in some detail, as are Georgesçu's noteworthy analyses of capitalist breakdown and overpopulation in agrarian economies.

Chapter 5 offers an "economist's primer on thermodynamics" to introduce the non-technical reader to some basics (and not-so-basics) of this branch of physics. Included are discussions of Georgesçu's views and their correctness, in addition to an analysis of the applicability of thermodynamic principles in economics.

Chapter 6 describes the role thermodynamics played in Georgesçu's economic work. Included is a discussion of the important problem of inter-generational resource allocation.

Chapter 7 describes "bioeconomics," a fitting way to summarize and integrate many of Georgesçu's views. Although Georgesçu-Roegen did not leave us anything that could be called a complete exposition of bioeconomics, his written descriptions are generally consistent. If one argues, as we do, that Georgesçu's purpose was to remake economics into a science useful for discussions of the survival of the human race, then bioeconomics offers Georgesçu's visions of what that science would be like. Environmental policy enters here, although Georgesçu felt we could make progress in this sphere only when people's *minds* changed. Changing opinions was seen by Georgesçu as a necessary prelude to changing policies, and he spent far more effort on the former task than the latter.

Because Georgesçu-Roegen believed that the inability of future generations to contract in current resource markets implied that current generations overused resources, he consistently advocated conservation, reduction in population levels, and a transformation of society toward primary reliance on organic agriculture. While many technologically optimistic economists argue that the great increases in yields obtained through modern (that is, chemical and capital-reliant) agriculture suggest that the limits to growth are not immediately important, Georgesçu criticized modern agricultural practices as energy squandering. Indeed, he saw the whole history of technological progress as a continuous substitution away from abundant sunlight toward increasingly scarce mineral resources. He felt that this process could not continue for long and that it was contrary to the long-run interests of the human race.

Although some modern ecologists and environmentalists argue their causes by attributing "rights" to animals or ecosystems, none of this sentiment is present in Georgesçu-Roegen's work. Rather, the end of all economic activity is human welfare: Man is the measure of all things. However, it is not solely the welfare of people alive today that is the issue.

While it is not easy to reach conclusions about the work of Nicholas Georgesçu-Roegen, we have reached several and they are summarized in our concluding chapter. We urge the reader to reach his or her own conclusions. Georgesçu-Roegen represents an important, though sometimes heretical, voice in twentieth-century economics and the philosophy of science. The issues his work raises are profound and deeply troubling. It is critical that those who profess to care about the future of Man become familiar with what he had to say.

NOTES

1. Screpanti and Zamagni (1993; 417).
2. Ibid., p. 417.

3. See, for example, the book review by Frank Adelman, a physicist, in the *Journal of Economic Literature*, (1972).
4. This is an important point, and we discuss it in detail in Chapters 5 and 6.
5. Georgesçu's cool reception by many economists is quite curious given the fact that, were he accepted, the importance of economists would be greatly elevated.
6. However, the Entropy Law does yield the "principle of the degradation of energy" and limits the amount of work obtainable from heat engines.
7. See Mayumi (1993).
8. Georgesçu-Roegen's most detailed bioeconomic model is presented in "Energy analysis and economic valuation" (1979).

2. Nicholas Georgesçu-Roegen: a scholarly refugee

The life of Nicholas Georgesçu-Roegen forms a bridge between many separated worlds. Born prior to the First World War, in a Romania of the "Old World," he died 88 years later in the "New." A witness to unimaginable atrocity and waste – two world wars, numerous dictatorships, the creation of the Iron Curtain – Georgesçu also knew and, in some significant cases, studied with, such scholars as Karl Pearson, Fréchet, Borel, A.C. Pigou, Joseph Schumpeter, Einstein, and many others. Never really a man of "action," Georgesçu served in numerous influential positions in the Romanian government between the late 1930s and 1946, and was, in several instances, a mere lucky accident away from arrest, torture, and death at the hands, in turn, of the fascist Iron Guard and the Communist puppets of the postwar era. A scholar through and through, short of stature, with a cherubic face and poor eyesight, Georgesçu-Roegen found himself at times negotiating over the fate of Romania with either Nazi officials or Soviet commissars: certainly not the sort of life one would have predicted for a small, precocious boy, with a flair for mathematics, from the small city of Constanta, Romania.

Among the most important, and most frequently cited, ideas in Georgesçu's scientific work is the notion of "hysteresis." This phenomenon, familiar to students of magnetism, refers to the impact of history, or external forces, on the nature of some property. By exposing a magnetic medium to an external electric force, it is found that, after the external stimulus is removed, the magnet will fail to return to its previous condition. Georgesçu felt – indeed, he directly *perceived* – that this type of phenomenon happened to people as well. It is fitting that in describing his life we take his word for this, and carefully seek to observe the hysteresis imposed on Nicholas Georgesçu-Roegen by his extraordinary experiences.[1]

Nicholas Georgesçu-Roegen was born in 1906 in the (then) small Black Sea port of Constanta, Romania. Georgesçu's family was of humble, but not peasant, origins, his mother a sewing teacher and his father an army officer. Georgesçu was always proud of the simple origins of his family, and expressed great affection for his mother, father, and brother throughout his life. Georgesçu's father died in 1914, having lost his army position, while his brother succumbed

9

to a fatal reaction to a vaccine given to Romanian soldiers destined for the war in 1941.

Georgesçu, around age 10, had the good fortune to be a student of an extraordinarily able teacher, Gheorghe Radolescu, who perhaps first interested Georgesçu in mathematics. Georgesçu frequently spoke of Radolescu in later years, and, in fact, dreamed of becoming a mathematics teacher throughout his youth and early adulthood, a testament to Radolescu's influence. Radolescu did two further services for the young "Nicholas St. Georgesçu", as he was then known. First, Georgesçu reports that Radolscu was one of only two instructors who ever boxed his ears. Second, Radolescu encouraged Georgesçu to apply for a scholarship at the prestigious Lycée "of the Monastery on the Hill," which he did, winning the last place scholarship in 1916. His installation at school, however, was delayed by Romania's entry into the First World War against the Central powers, necessitating Georgesçu's first (but not last) sojourn as a refugee. His family fled to Bucharest and, in considerable hardship, they lived with his maternal grandmother. During this period, Georgesçu had his initial direct encounter with the consequences of war when he was traumatized by the sight of bloody bodies piled up in a city street.

In 1918, Georgesçu, now 12 years old, returned to the Lycée, and began an extremely rigorous course of classes and physical work under the direction of an apparently outstanding faculty that included, among others, the probabilist Octav Onicescu. Georgesçu was an excellent student, particularly in mathematics, and he took prizes for small publications in the math newsletter *Gazeta Mathematica* in 1923 and 1924.

After graduation from the Lycée, Georgesçu was accepted at Bucharest University for study in mathematics. In case the reader is wondering how a boy from a poor family could pursue such an education, it is important to note that education in Romania was even then state-sponsored and entirely free to students who qualified for admission.

Georgesçu did very well at Bucharest, mastering the rather old-fashioned maths curriculum then in force. While he remained unexposed to topology, set theory, and similar "modern" topics, he did benefit directly from a course on singularities of differential equations, and he used some of this knowledge to construct his famous arguments on integrability in "The pure theory of consumer's behavior" (1936). There is no evidence that Georgesçu studied economics at Bucharest, although statistics did begin to attract his interest. Further, at this time he obtained extensive insight into Romanian peasant society through supplementing his extremely meager resources by teaching mathematics at a "peasant lycée," a government school in the countryside.

Georgesçu graduated from the University of Bucharest with an honors degree in mathematics in 1926. A degree was not all Bucharest gave him, however: as a student he met Otilia Busuioc, who became his beloved wife soon thereafter.

Also a mathematician of some ability, Otilia became Georgesçu's soulmate, and they remained inseparable until his death in 1994.

Georgesçu, on the recommendations of his professors, obtained a grant allowing him to study for the PhD at the University of Paris (Sorbonne) in 1927 and, suitcase in hand, he left for France in November 1927. He almost failed to make it: political favoritism by officials of the National Peasantist Party (NPP), which he later joined, almost deprived him of the award.

Georgesçu in later years would often reminisce on his life in Paris at the Sorbonne. This was a time of both extreme physical hardship and intellectual exhilaration. On the one hand, he was extremely poor, sometimes even hungry. His meager stipend from Romania, supplemented on occasion by even more meager remittances from his family, left little margin for indulgence. Compounding his physical discomfort was rather disgraceful abuse he and other *metèques* students endured from their French compatriots.

In intellectual terms, however, Georgesçu's Paris days were extremely rich. For example, he was able to take classes from Lebesque, Fréchet, Borel, and other great mathematicians, and he knew François Divisia, of index number fame, quite well. Georgesçu, prior to the Second World War, had a remarkable knack for being in the right place at the right time, and his Paris experience certainly bears this out.

Georgesçu-Roegen's dissertation title, loosely translated, was "On the problem of discovering the cyclical components of a phenomenon." Georgesçu proposed a new and superior method in "periodogram" analysis that allowed coefficients to be found for certain types of random processes having trigonometric representations. In particular, he examined rather fancy cosine expansions for periodic series of random variables exhibiting both drifts and recurring components. His method, which he accurately felt was ahead of its time, was summarized in an article in *Econometrica* in 1948. Schumpeter actually used Georgesçu's technique, in a simplified form, in his monograph *Business Cycles* (1939).

Georgesçu's dissertation was accepted and he was awarded his doctorate "*avec les félicitations du jury*", an extraordinary honor. His interest in statistics, which he never lost, caused him in 1930 to move to London, where he studied with the great statistician Karl Pearson. During this time, Georgesçu lodged with a poor family, the Hursts, from whom he learned to speak English. Also at this time he had his first encounter with two important academic institutions: circulating libraries and tuition. Despite these novel experiences, Georgesçu managed to spend a year working on the "problem of moments", one of the deepest issues in statistics. Pearson pioneered research on this problem, although he did not "solve" it. Neither did Georgesçu.[2]

In 1931, Georgesçu was invited to apply for a Rockefeller Foundation grant to study in the US. He was quite interested and, as part of his application, expressed his desire to see his periodogram analysis applied to the (to him) "famous" Harvard Economic Barometer, an early business cycle forecasting project. After returning to Romania, he set off for the US in 1933. In addition to his journey to the New World, he adopted his new name, Nicholas Georgeşçu-Roegen (rather than the earlier Nicholas St. Georgeşçu) on his first book, a statistics text titled *Metoda Statistica*.

Georgeşçu arrived in New York, made his way to Cambridge, Massachusetts, and found rooms near Harvard. His search for that Holy Grail, the Harvard Economic Barometer, ended in disappointment when he learned it had failed to predict the Great Depression and had been ended in late 1929. Somewhat as a Plan B, he went to see the economic-cycle researcher, a Professor Joseph Schumpeter. That fateful meeting was eventually to change the entire direction of Georgeşçu's work.

Georgeşçu was treated most kindly by Schumpeter, and soon met another influential friend, Wassily Leontief. As his interest in economics, stemming mostly from economic statistics, grew, Georgeşçu was invited to participate in Schumpeter's now famous "circle" for economic discussion. These meetings, which must have been extraordinary, included at various times Oskar Lange, Paul Sweezy, Leontief, Gerhard Tintner, and Fritz Machlup, in addition to Schumpeter as "ringmaster." This experience was the basis of Georgeşçu's training in economics; he never formally enrolled in any economics classes. He would sometimes joke about this in later years, offering that he attended only "Schumpeter University."

In 1935, Georgeşçu obtained a modest stipend from the Rockefeller Foundation to support travel about the US. Somewhat reluctantly, he and Otilia began an extended road trip about the country, journeying as far as California. Georgeşçu exploited his opportunities and managed to speak with such figures as Henry Schultz, Harold Hotelling, and Irving Fisher. Further, Georgeşçu visited the Cowles Foundation, then in Colorado, and enjoyed a meeting with Albert Einstein at Princeton. To his later regret, Georgeşçu failed to ask Einstein about the Entropy Law, so it may safely be concluded that such concerns were not yet on his mind in the mid-1930s.

The middle of the depression decade was a time of great achievement for Nicholas Georgeşçu-Roegen, and his publications during this interval established his reputation in mathematical economics. His first publication in economics was "Note on a proposition by Pareto" in the *Quarterly Journal of Economics* (*QJE*) in August 1935. This was soon followed by a paper in the *Review of Economic Studies* (October 1935) (on marginal productivity in Leontief-type models), an article on the "constancy" of the marginal utility of income (*QJE*, May 1936), and the famous "The pure theory of consumer's behavior" (*QJE*, August 1936). Georgeşçu's work on the marginal utility of income, though now superseded, was for its time very modern in spirit. He revived this theme many years later in "Revisiting Marshall's constancy of the marginal utility of money" (*Southern Economic Journal*, 1968). The interested reader is urged to compare these two papers, published 30 years apart.

Despite earning the admiration of Schumpeter, the support of Harvard, and an enviable start in mathematical economics, Georgesçu and Otilia returned to Romania in 1937. This decision arose primarily from patriotism and Georgesçu's gratitude for the educational support given him by Romania during his studies in Paris and London. This decision was one that he would forever regret. It began, he would often say, his "exile," and in retrospect exile is perhaps an apt term. Although Georgesçu served as a professor at Bucharest and began his government career, a review of his academic work for the period 1938–46 shows an almost total absence of economic research. Instead, sporadic publications in Romanian on history and trade predominate. Given the circumstances of these years it is miraculous that he wrote anything.

On his return trip to Bucharest, Georgesçu did manage to visit London, where he heard Hayek lecture at the London School of Economics on monetary phenomena, an area Georgesçu freely admitted he never felt comfortable with and on which he published very little. This journey also introduced him to R.G.D. Allen (whom he had criticized in print for mathematical errors), and Sir John Hicks, who became his friend.

Georgesçu's life in Bucharest was eventful and, for him, enormously frustrating. As a "doctor" from the Sorbonne, he was regarded as something of a sage in a country that exhibited surprisingly high illiteracy rates. As a result, Georgesçu was tapped for numerous government posts, for some of which his training was useless. However, he directed the Central Statistical Institute in Bucharest, a government office that compiled both economic and demographic statistics, and worked on the *Enciclopedia Romaniei*.[3] He subsequently served on the national Board of Trade and directed negotiating teams charged with making trade arrangements with Germany, the USSR, and others.

The Romania to which Georgesçu returned was a violent place of instability and hardship. The Romanian fascists, called the "Iron Guard", were actively seeking power through assassination and political terror. The prime minister Calinescu was killed by fascists in late 1939. The king, Carol II, tried to suppress the Iron Guard but was eventually forced out by them. The West was ineffective in aiding Romanian resistance to Stalin, a miscalculation that led to the appointment of a pro-German premier after Calinescu's death. By 1941, pogroms against Jews and political opponents of the Iron Guard were underway. The world's darkest nightmare was at hand.

Georgesçu in the late 1930s had joined the National Peasantist Party (NPP), which favored land reform along democratic paths. An opponent of both fascist and communist/Stalinist ideologies, Georgesçu found his position quite dangerous. He continued to serve as a technical bureaucrat in the government, and perhaps his academic standing protected him somewhat. He narrowly escaped arrest or death on several occasions.

By 1944, Romania, an Axis power, was losing the war and continual bombing of Bucharest was common. The arrival of the Soviet army, by Georgescu's account an undisciplined mob, made matters worse and he fled Bucharest, joining his mother and wife in the countryside.

As an opposition party, the NPP saw its stock rise somewhat with the defeat of the fascists, and Georgescu found himself appointed head of the Romanian Armistice Commission, a very important post that involved negotiations with the Red Army. Although he hoped this would be an opportunity to secure better terms for Romania, the Soviets viewed the Commission as a mechanism for collecting heavy war reparations levied on the Romanians. This could not have been a very pleasant job.

In 1946, the communists, backed by the USSR, stole the national elections and established a puppet government in Bucharest. Georgescu found himself jobless, nearly destitute, and in some danger of arrest and even execution. The abdication of Carol's son in 1948 signaled the end. Georgescu and Otilia decided to escape.

They managed to obtain counterfeit travel documents from a Jewish underground group. Hiding out in Constanta, Georgescu negotiated for passage aboard the freighter *Kaplan*, bound for the Bosporus. Hiding in a small crate, Georgescu and Otilia successfully evaded a search and escaped to Turkey. Having previously traveled to Turkey on official business, Georgescu was able to obtain financial help from friends and soon contacted Harvard. Offered a job in Cambridge, Georgescu began seeking a way to the US.

Travel of any sort in Europe was naturally quite difficult in the immediate postwar years, and Georgescu and Otilia had to overcome many difficulties. However, fate smiled on them, for they were eventually able to obtain the transit passes required to leave for the US, arriving in late 1948. Georgescu obtained employment as a lecturer and research associate at Harvard University, a result of his previous work and connections at that institution.

Although Georgescu believed – probably correctly – that Harvard would eventually have extended him a permanent faculty position, the tremendous insecurity of the war years allowed George Stocking, the famous institutionalist economist, to lure Georgescu to Vanderbilt University with a permanent academic position. Thus, in 1949, Georgescu began his employment at Vanderbilt, continuing there (except for numerous visiting appointments around the world) until his retirement at age 70 in 1976.[4]

At least one commentator has attributed Georgescu's "isolation" within the economics profession to his long employment at Vanderbilt. Certainly, Nashville was neither Cambridge nor Chicago nor New York. Further, had he remained at Harvard, it is probable that his influence on economic theory would have been greater, for surely his graduate students would have been, on average, better. However, the Vanderbilt economics department which Georgesçu joined was neither undistinguished nor inhospitable to him. First, in addition to Stocking (who served as President of the American Economic Association in 1958), William ("Bill") Nichols, acclaimed for his empirical work in industrial economics, taught at Vanderbilt.[5] Second, Vanderbilt was internationally known as a focal point for the "Agrarian Movement" among southern American writers and intellectuals. This tradition impressed Georgesçu, and he mentioned the Agrarians frequently in his writings.[6]

Finally, and perhaps most importantly, Georgesçu found Vanderbilt a tolerant and supportive environment. As is frequently the case with brilliant scholars, Nicholas Georgesçu-Roegen suffered from considerable feelings of insecurity, a state probably exacerbated by his precarious existence in Romania during the war. These insecurities sometimes made him a difficult colleague.

Georgesçu made no secret of his feeling that much of his work had been ignored. This had one curious effect: in order not to make the same mistake by ignoring the writings of others, he made a habit of citing very early writings in many of his papers. For example, one frequently finds citations to Marshall, Pigou, Pareto, Gossens, and others of the same antiquity, even in theoretical papers written in the 1960s.

Georgesçu-Roegen ordinarily taught only graduate classes at Vanderbilt, his focus being economic theory, microeconomics core courses, and occasionally statistical and mathematical economics. The contents of Georgesçu's microeconomics graduate course is revealing, both personally and professionally. First, an extensive study of consumer theory was undertaken and much emphasis was placed on fundamental issues, such as the existence of indifference curve representations of preferences. Revealed preference analysis was also highlighted. Production theory also received attention, and here we find something interesting. As early as 1960, Georgesçu carefully integrated his production lectures with references to both *time*, and its role in production, and the entropic nature of materials processing, the latter playfully characterized as producing "devil's dust." The idea here is that, through economic activity, valuable material substances are scattered widely so as to become irrecoverable. For example, the amount of any mineral on earth is, save for the needs of spacecraft, virtually fixed. Yet we can contemplate "shortages" of strategic materials. The reason, of course, is that economic use of these materials results in their being widely scattered, at potentially very low concentration, throughout the environment. Georgesçu felt this idea sufficiently important to include in his graduate microeconomic classes in the late 1950s.

His intimate, first-hand knowledge of source material in the history of economic thought also influenced his curriculum choices.[7] Economists receiving much attention in his theory courses included Schumpeter (particularly on development themes), Leontief (input/output models), Marx, and, interestingly, Quesnay (of *Tableau économique* fame). Indeed, many of Georgesçu's writings on political subjects (such as his discussion of "elites" in *The Entropy Law*, 1971) and the sequential epistemological framework of production arising from consideration of entropy changes, bear a resemblance to the analyses of the Physiocrats such as Quesnay (1694–1774) and Turgot (1727–81).[8]

Georgesçu often used no assigned texts in his classes (telling one student: "Don't read anything – it will only confuse you"), and relied to a far greater degree on tutorials, in the European sense, than is ordinarily the case in US universities. As might be expected from an inspection of his publication record, Georgesçu himself was tireless, often continuing his work at home in the evenings and laboring late into the night. He rarely discussed his current work with other economists or students, and collaborated in very few projects over his career.[9]

Georgesçu held numerous visiting appointments and research fellowships between 1950 and his retirement in 1976. He was Rockefeller Visiting Professor at Osaka University (1962), a lecturer at the Indian Statistical Institute (1963), a Ford Fellow lecturer or visitor in Brazil several times in the 1960s and early 1970s, and visiting professor at Florence (1974), Ottawa (1975), West Virginia (1976–78), University Louis Pasteur (Strasbourg) 1977, the Technical University of Vienna (1978), Texas (1979), McGill (1979), and others.

Georgesçu received many academic honors throughout his career. Besides his early Rockefeller Foundation Fellowship (1934–36) for study in the US, he was a Guggenheim Foundation Fellow (1959), Fellow of the International Institute of Sociology (1960), Fellow of the Econometric Society (1950), Distinguished Fellow of the American Economic Association (1971), the Richard T. Ely lecturer (1969), and Distinguished Associate of the Atlantic Economic Association (1979). In addition, he won the prestigious Earl Sutherland Prize for Research in 1976. Late in his life, Georgesçu was inducted into the Romanian Academy, which pleased him very much.

Georgesçu served the economics profession in several editorial capacities over his career. First, his work as Associate Editor of *Econometrica* (1951–68) has been the subject of some recent research.[10] The latter years of his tenure at *Econometrica* must have been rather strained, given the growing iconoclasticism of his views on econometrics and the proper application of mathematics in economic research.[11] Georgesçu also served on the editorial boards of *Fundamenta Scientiae* and *Metroeconomica* in his later years.

Georgesçu received honorary degrees from University Louis Pasteur, Strasbourg (1976),[12] the University of Florence (1980), and the University of the South (US, 1983), and was honored by his long-time employer, Vanderbilt University, with the Harvie Branscomb Award in 1967, a very significant honor.

Georgesçu-Roegen was, by any standard, a remarkable man. The author of perhaps 200 academic journal articles, several books, and many reports, the solver of the integrability puzzle, the discoverer of the non-substitution theorem, a father of environmental economics, Georgesçu's record of academic achievement is extraordinary. Yet it is undeniable that Georgesçu's impact on the economics profession is far less than any uncolored reading of his record might suggest. Although it is not the purpose of this narrative to offer "sociological" explanations for this state of affairs, it is important at least to catalogue some possible causes.

First, as mentioned above, Georgesçu pursued most of his career at Vanderbilt University. As a consequence, the graduate students he was able to teach were not, on average, as good as those he might have had at either Harvard or some other leading department. Georgesçu periodically expressed disappointment at his students, and emphasized the importance of cultivating the occasional "diamond in the rough." To the degree that an economist, possessed of an original and independent vision, must rely on his or her students to establish that vision in the mainstream, Georgesçu's various "exiles" almost surely reduced the impact of his work. On the other hand, Vanderbilt did afford him great freedom and accommodated some of his eccentricities, so the net effect is unclear. Further, Georgesçu's students included Herman Daly, Kozo Mayumi, and others, so it is difficult to argue that his teachings lacked for effective expositors.[13]

It must also be recognized that the lack of collaboration by Georgesçu-Roegen was no coincidence but, at least in part, was a result of the fact that he would have been a difficult collaborator. His sometimes inflexible – even aggressive – approach was surely, at times, a handicap. For example, Georgesçu, a Distinguished Fellow of the American Economic Association, resigned from the AEA as a protest at the *American Economic Review*'s publication of a landmark study examining the economic content of animal behavior.[14] Most economists, even if unsympathetic with animal studies, would probably not follow so extreme a course.

Georgesçu was aggressively intolerant of mistaken economic ideas, and this sometimes translated into intolerance of the mistaken economists. Some economists who found themselves on the receiving end of these criticisms were often sympathetic to his work. On the other hand, Georgesçu could be gracious towards those he felt were right. No doubt these feelings arose from his belief that economics "is a serious subject."

Georgesçu's very difficult literary style also contributed to his isolation among economists. He rarely summarized his arguments, even when such summaries could have greatly aided the reader. This, combined with his difficult temperment, had unfortunate effects. Georgesçu's autobiographical writings refer to a general feeling of ostracism. In Szenberg (1992), Georgesçu remarks:

> If I finally realized that I was running against one current or another, it was not from any crossing of intellectual swords with my fellow economists, who have systematically shunned such an encounter, but from their personal attitudes towards me.

Georgeşçu attributes these personal attitudes to his economic positions, particularly his beliefs in the inefficiency of marginal pricing in agrarian economies, his criticisms of analytic methods in economics, and so on. It is a tribute to Georgeşçu's love of scholarship that he could accept such an explanation. He felt slights quite keenly: he mentioned that his 1969 Ely lecture was scheduled at the same hour as the meeting of Fellows of the Econometric Society. Many similar anecdotes are available.

Certainly, Georgeşçu's work took a heterodox path, and the lot of the heretic is never an easy one. However, it is probably also fair to say that he has not always been well served by his "followers." Because of his obscure literary style, many readers, including prominent neoclassical economists, found some of his work impossible to fully understand. Due to his application of principles from physics to economics, many writers with extreme views have been attracted to Georgeşçu, each finding in his obscurer work whatever they wished. Other "heterodox" economists, such as von Mises and Veblen, have suffered similar treatments by some of their supporters. The apparent lesson is that, if one is both obscure enough and profound enough, the risk of cooption is large.

Although Georgeşçu certainly made several startlingly "radical" policy proposals (for example, the Dai Dong Association proposals to "globalize" the ownership of resources and abolish immigration controls), many readers may be startled to learn that he was regarded by his friends as basically a political conservative (in the twentieth-century US sense), almost a Republican.[15] Yet there was clearly an element of elitism, or *noblesse oblige*, in Georgeşçu's outlook. The poor of the world were the primary concern and their welfare the primary aim. In this we see the basic patriotism of a poor boy from Romania, grateful for his educational opportunities, never forgetting his experiences in eastern Europe, and using his intellectual powers to improve the lot of those whose work made his education and, ultimately, his escape possible.

That Georgeşçu's last years were marred by bitterness has been remarked on by others.[16] Undoubtedly, a feeling that much of his work had been ignored contributed greatly to this outcome. Of course, as a famous economist, Fellow of the AEA and Econometric Society, and so on, it is hard to accept any characterization of Georgeşçu's career as a failure. However, one rarely sees oneself in an accurate light, and probably he was no exception to this rule.

Nicholas Georgeşçu-Roegen died in Nashville, Tennessee, on October 30, 1994, at the age of 88. Otilia died in Nashville in early 1999. They had no children.

NOTES

1. Sources for this chapter include Georgesçu's autobiographical writings, including the series, "Emigrant from a developing country," his contribution to Szenberg (1992), recollections of the authors, and interviews with long-time colleagues including Elton Hinshaw, V. Kerry Smith, and John Siegfried. The reader is also urged to read H. Daly's obituary essay (1995).
2. The "problem of moments" asks the question: "given the infinite but denumerable list of moments of a distribution, can one find the density function?"
3. Some drafts of this encyclopedia were apparently destroyed by the communists in 1945. Georgesçu does list it on his *vita*, however. We are unable to locate any published version.
4. Georgesçu was granted an extension of the mandatory retirement age of 65 and did not have to retire in 1971. The volume *Evolution, Welfare, and Time in Economics: Essays in Honor of Nicolas Georgesçu-Roegen* (Tang *et al.*, 1976), honored his retirement.
5. Georgesçu had a very high regard for both Stocking and Nichols.
6. Georgesçu did not praise their economics, however. See "Process in farming versus process in manufacturing" in Papi and Nunn (1969).
7. Georgesçu was sometimes described as reading "nothing written after 1960". His primary references for topics in the natural sciences are even older, often truncating around 1930.
8. The Physiocrat's notion of "unproductive labor," the emphasis on timing and sequence in agriculture, and the idea that only agriculture and mining can produce a "surplus," all resemble ideas in Georgesçu's work.
9. An exception was in his early statistical work in genetics, "L'Influence de l'age maternal, du rang de naissance, et de l'ordre des naissances sur la mortinalité" (1937), with R. Turpin and A. Caratzali.
10. See Gayer (1997). Georgesçu's correspondence is archived at Duke University, in the Economists' Papers Project. See Weintraub *et al.* (1998).
11. See, for example, Georgesçu's account of a dispute with Malinvaud and J. Dréze in Szenberg (1992).
12. Strasbourg hosted an international conference on Georgesçu's scientific work in November 1998.
13. This is not to suggest that either Daly or Mayumi hold identical views to Georgesçu. In fact, areas of strong disagreement exist in some cases.
14. Georgesçu mentions this episode in Szenberg (1992).
15. The Dai Dong Association, a creation of the Fellowship for Reconciliation, brings together scientists and scholars from around the world to study war, the environment, and global poverty issues. Georgesçu participated in Dai Dong at the famous Stockholm Conference of 1972.
16. See Daly (1995).

3. Georgesçu-Roegen's epistemology and economic methodology

3.1 INTRODUCTION

A distinguishing characteristic of Nicholas Georgesçu-Roegen's economic research, evident even when he dealt with "mainstream" problems, is a deep and unwavering concern for economic methodology and epistemology. Even in his earliest work in mathematical economics, such as his celebrated "The Pure theory of consumer's behavior" (1936), Georgesçu's writings are characterized by a strong philosophic impulse, leading him to minute examinations of the logical underpinnings of economic science. A part of this legacy arose, no doubt, from his most unconventional educational background, and the roles played by economic and scientific methodologists such as Schumpeter and Pearson in his early training. The stylings of the scientific journals of those days allowed him to indulge his passion for philosophy in print to a degree unlikely to be repeated in our times. Thus, we enjoy in Georgesçu's work an opportunity to observe the development of a powerful methodological position as economics evolved into its "modern" form over his lifetime. That Georgesçu had an important role in the development of modern economics increases our interest in his epistemological speculations. That his views ultimately form a powerful indictment of modern economics is, therefore, especially significant.

Although it is difficult to select from among Georgesçu's vast and, at times, apparently inconsistent pronouncements on philosophical matters, a modest list of elements appear repeatedly in his work and represent themes that, to a degree, inform his work in all areas of economics. Before beginning a systematic review and integration of these ideas, it is important to provide some background. Georgesçu, as with all of us, was partially a product of his times and experiences.

Perhaps the most relevant aspects of Georgesçu's life for his later philosophical speculations include his extensive training in mathematics and statistics, his close work with Schumpeter and Karl Pearson, his early life of relative poverty in the surprisingly cosmopolitan environment of Constanta around the First World War, and his dreadful experiences during the Second World War.[1] Broadly drawn, these experiences appear to have established the following pattern in Georgesçu's work. First, he was quite conversant with higher mathematics and was convinced that fundamental ideas in mathematics, such as transfinite cardinal arithmetic, were relevant to economics, though not in an immediately practical way. This conclusion is well illustrated by his emphasis on measurability, change, dialectics, and evolution.[2] Second, institutional detail was, for Georgesçu, an essential aspect of any useful model. Having lived in Romania, the US, and Europe, and having traveled extensively in South America and other areas, he was convinced that economic activity and institutions were sufficiently varied to require analyses by different sorts of models. In particular, Georgesçu regarded most familiar neoclassical models as applicable only to "advanced" economies for which issues of subsistence were irrelevant.

Georgesçu's experiences living under war and dictatorship and his experiences as a refugee also affected his methodological outlook. It is fashionable nowadays to regard (neoclassical) economics as a body of knowledge useful for positivist calculation: moral issues are not amendable to economic analysis, although many actions which might be regarded as having a moral component (such as drug addiction) are analyzed by economic models. Georgesçu saw this trend as rather strange. The purpose of economics is to inform our discussions of the "best aims of mankind." A deep humanism and an abhorrence of the wastefulness of war in both the material and spiritual dimensions are characteristics of Georgesçu's work.

His methodology, though complex, contains a set of standard themes that, at the risk of offending students of his work who will find our list incomplete, we may profitably summarize. Coarsely put, the main epistemological conclusions of Georgesçu-Roegen are:

1. The set of real numbers is an entity built from the unions of sets of discretely distinct elements, and thus cannot correspond to the continuum of our experience.
2. All sciences, except mathematics and certain branches of physics, propose laws that involve relationships between "quantified qualities," and the hallmark of such laws is non-linearity. Laws between strongly cardinal magnitudes are linearly homogeneous.
3. Neoclassical analysis in economics confuses the ideas of order and measure.
4. Many phenomena of economic significance, such as the development of economic institutions, cannot be understood, nor even defined, by atomistic investigations that seek to establish the properties of wholes only by reference to properties of the constituent parts. Societies, as well as species, undergo evolutionary development.
5. Important economic categories, including such concepts as "commodity," "expectation," and "money," are dialectical rather than arithmomorphic

    ideas, and do not exhibit the discrete boundaries necessary to adequately represent them by real numbers.

6. Models that are constructed from ordinary functions on real numbers are termed arithmomorphic, and such models are mechanical analogies.
7. The economic process cannot be adequately represented by mechanical analogies, because such analogies cannot represent any evolutionary process.
8. Man's economic activities are an extension of his biological struggle for existence. Any meaningful analysis of economic activity at the macroeconomic level must accord a unique role to natural resources. Methodologically, this can be done by including in the model stocks that either are depleted or accumulate through time. The resulting analysis, while only an analytic simile, will be an improvement over current models.
9. The Entropy Law constitutes an important example of a constraint on economic activity that has no arithmomorphic representation.

The balance of this chapter outlines these conclusions in greater detail, and summarizes the sources of Georgesçu's beliefs.

3.2   MATHEMATICS AND DIALECTICS

Georgesçu and the Continuum: Some Background

Many of the points Georgesçu makes about change, evolution, and scientific method are unintelligible to readers without at least a passing acquaintance with several concepts from (naive) set theory, transfinite arithmetic, and related issues in analysis and measure theory. What follows is intended only as the briefest of informal introductions to these difficult topics. The interested reader is urged to consult standard texts, such as Kaplansky (1977), for a more detailed view. Readers with substantial prior knowledge in these areas may profitably skip this introduction entirely.

    That Georgesçu felt that an understanding of set theory and related topics was important is obvious from several pieces of evidence. First, *The Entropy Law and the Economic Process* (1971) includes a long appendix on the "Structure of the Arithmetic continuum." Second, the ideas embodied in that appendix are evident in numerous important papers by Georgesçu, including those on consumer preferences, on production, and on "measurability." Third, a failure by many economists to understand these issues produced the "ordinalist's fallacy," the novel basis for Georgesçu's criticisms of expected utility theory and other building blocks of neoclassical economics.[3]

    We begin with the natural numbers, N, comprised of the set {1, 2, 3, ...}. These most familiar of all numbers have several properties. First, they are cardinal numbers, which means to Georgesçu that they satisfy certain conditions representing the *mechanical* (or physical) operations of addition and subtraction. Being cardinal, they are devoid of quality in the sense of "indifferent subsumption": 4 minus 2 equals 2 regardless of which "two" one "takes away."

Defining operations on N can lead to a problem. First, while $a \in$ N, $b \in$ N implies $a + b \in$ N, subtraction quickly necessitates the notions of negative numbers and zero. The resulting set of integers, I, itself has "holes" in it. Division, in particular, creates fractions a/b where $a \in$ I, $b \in$ I, $b \neq 0$, and $a/b \notin$ I. These numbers arise in measurement and in ancient times it was thought that this set of *rational* numbers constituted all numbers.

The discovery of the irrational numbers must have been one of the most upsetting developments in the history of humanity.[4] However, simple plane figures (for example, squares) with integral length sides quickly give rise to irrationals such as $\sqrt{2}$. Such numbers cannot be written as a ratio of integers. Thus, an entirely new class of numbers was added. Surely, now we had found all the numbers there were.

As Georgesçu noted, more and more numbers were invented/found as the need arose for them. The integers, the rationals, the algebraic numbers and, finally, the reals were described by mathematicians. The set of real numbers R constitutes all numbers that can be written as infinite decimal fractions, and R contains all numbers familiar to most readers.

The collection of reals R is often visualized as the real "line," a dense collection of points, each point a real number, seemingly gapless and continuously extended. Of course, history suggests that optimism over the "completeness" of this "arithmetic continuum" may be misplaced. Nevertheless, the properties of the continuum are fairly well understood, and most economists certainly use the reals without thinking that they might be, in some important sense, inadequate.

A number of elementary set-theoretic propositions can usefully be introduced. First, there is an important distinction between cardinal and ordinal numbers. In Georgesçu's common-sense view, cardinal numbers, or, more generally, cardinal variables, are objects on which the intuitive analogy of certain physical operations can be performed. These operations roughly correspond to addition and subtraction, with a recognition that the constituent units of the magnitudes are free of any qualitative variation.

Thus, if the quantity of water in a vessel is equal to 1 gallon, and we take away one-half of this amount, then one-half gallon remains regardless of which half gallon we remove.

Ordinal numbers, on the other hand, order things only, and no notion of measure is implied. Hence, it is not meaningful to "add" ordinal numbers. This point is, of course, well known to economists from modern textbook treatments of utility theory and utility functions. Utility (or "ophelimity") is merely the assignment of elements of R to various "bundles" or outcomes in order to represent their ranking. Objects which may be completely ranked by labeling them with real numbers are called "ordinally measurable," although perhaps "orderable" is a better nomenclature.

Given the common sets of numbers familiar to most students, N, I, R, and so on, one naturally wonders whether and in what sense one can say one set is larger than another. The great mathematician Georg Cantor (1845–1918) solved this and many related problems by proposing a theory of infinite sets called transfinite arithmetic.[5]

Cantor's basic insight was that two sets, A and B, whether infinite or finite, have the same number of elements if and only if all their elements may be placed in a one-to-one correspondence. Thus, if every member of A may be uniquely paired with a member of B, and vice versa, then A and B have the same number of elements. We call this (common) number of elements the cardinal number of A (or B). One must be careful, however, with this definition: one must pair all elements of A with elements in B and vice versa.

The cardinal number of a finite set is merely the number of elements the set contains. The cardinal number of an infinite set, however, is called a transfinite cardinal number. It is easy to show that the sets N, I, the even integers $\{2, 4, 6, 8...\}$ and the primes $\{3, 5, 7, 11 ...\}$, all have the same cardinal number, denoted here α_0 ("aleph null").[6] Less obvious are the conclusions, proved in most undergraduate courses in set theory, that the rationals $\{a/b\}$, $a \in$ I, $b \in$ I, $b \neq 0$, and the algebraic numbers (roots of finite polynomial equations with rational (equivalently, integer) coefficients) also have cardinal number α_0.

The arithmetic continuum R contains both the algebraic and transcendental (not algebraic) numbers (for example, π), so its cardinal number cannot be smaller than α_0. In fact, it is very much larger. The cardinal number of the continuum we will denote by c, and c is larger than α_0: $c > \alpha_0$. This result was proved by Cantor using his famous diagonalization technique to show that the reals were not denumerable (that is, could not be counted by placing them in correspondence with N). Hence, there are "many more" real numbers than there are integers.

Curiously, the cardinal number of the real plane is the same as that of the real line. All finite dimensional Euclidean spaces have cardinality c. Despite this, there cannot be any *continuous* mapping between, say, the line and the plane. (There are "weird" mappings from the line to the plane that are not one-to-one, however). Thus, the fact that two sets have the same cardinal number does not imply that continuous maps between them are possible. This fact is actually used by Georgesçu in connection with the ordinalist fallacy and lexicographic types of preferences, as we will see below.

Since the time of Zeno, men have wondered whether time and space (or R) consisted of tiny, discrete "pieces." Modern mathematics examines this question using the notion of "set ordering."

Any set A is called "ordered" if there is a relation, denoted by a symbol such as <, such that, for any elements of A, we have (a) if $a \neq b$, then either $a < b$ or $b < a$; and (b) if $a < b$ and $b < c$, then $a < c$ (transitivity). Read the symbol < as "precedes". Any subset of an ordered set is ordered, and clearly any ordered set contains at most one first and one last element. Obviously, ordering of some sort is necessary to count set elements or otherwise pair them together.

A set is called "well-ordered" if it is ordered and all of its subsets, including itself, has a first element under its order relation. (We regard the empty set as well-ordered.) Thus, the closed unit interval [0, 1] is not well-ordered under the "less than" order because its open subset (0, 1) has no first element. Subsets of well-ordered sets are also well-ordered. Georg Cantor argued that every set can be well-ordered, an intuitive claim. Zermelo proved this claim to be true, although to do so he made use of a special assumption, the Axiom of Choice. There are very many equivalent formulations of this axiom, a transparent one being: "Given any set of mutually disjoint non-empty sets, there is some other set that contains a single member from each of these sets." This "innocuous" claim is actually equivalent to several important propositions: (a) for any two cardinal numbers d and e, either $d > e$, $d = e$, or $d < e$; (b) Zorn's Lemma: every non-empty partially ordered set A in which each chain has an upper bound that is in A must have a minimal element; and (c) the product of compact spaces is compact given the product topology (Tychnoff Theorem).

Without the Axiom of Choice (or its equivalent), in general, many basic propositions in mathematics cannot be shown, such as the result that the union of a countable set of countable sets is itself countable. Although there are alternatives to the Axiom of Choice, it is probably true to say that use of the Axiom of Choice is widespread in mathematics and, where applicable, in abstract mathematical economics.

The implications of the Axiom of Choice are sometimes intuitively appealing and at other times very strange. To see the strange aspect, note that any well-ordered set S must contain a "smallest" element. Call it e_1. But then, in the complement of the singleton set $\{e_1\}$, there is a "smallest" element e_2, and so on, leading to a sequence of the form $e_1 < e_2 < e_3 < ...$ Clearly, this sequence may be infinite. This sequence may not contain all the members of S but, if it doesn't, we may begin again with that part of S which is left. In this way we "eventually" exhaust S. Every sequence then has a "smallest" (that is, first) element. If the Axiom of Choice is accepted, then the real line can be well-ordered and has a "smallest" element. How one is supposed to do this is problematic.

The significance of these considerations for Georgesçu's epistemology is fairly immediate. In particular, given the Axiom of Choice, it is found that the real line is, apparently, "granular," "beads on a string" *without* the string, in Georgesçu's words.[7] Thus, for Georgesçu, the essential characteristic of a real number is its discrete distinctness: the arithmetic continuum does not correspond to the intuitive continuum of human cognition. Our notions of the continuum arise, for Georgesçu, from our direct experience of Time's flow. Yet, Time is *not* a dense sequence of "nows," just as motion is not a dense sequence of "rests." The arithmetic continuum, although a great scientific achievement in both conception and application, is not an adequate palette with which to portray becoming, real change, or evolution.

Some of these issues are illustrated by considering Zeno's paradox. To recall, Zeno's paradox (or, more properly, paradoxes, since there are several) aims to establish the implausibility of locomotion, the very simplest sort of change considered by science. In particular, the parable of the arrow notes that, for an arrow to move from A to B, $A \neq B$, it must move through infinitely many discretely distinct intermediate states. This is claimed to be impossible. This paradox does not show, of course, that motion is impossible; instead, it points to a problem that by necessity arises from the arithmetical conception of space (and time). The concepts of rest or motion are meaningless "at an instant." A still picture of a train cannot reveal if it is moving or at rest. Motion is a happening, if a very simple one, so it requires time. Georgesçu exhibited his great interest in Zeno's paradox in many of his writings and, contrary to what one might think, he saw the paradox as actually paradoxical and important.

The reader may object, of course, that science successfully treats the problem of locomotion by identifying a correspondence between a time variable (a real number) and a space/location variable (also a real number).[8] Motion is then identified with this function linking time and space. Such a view, Georgesçu admits, is adequate in many scientific applications yet, this mathematization of the problem is not really an answer to Georgesçu's complaints. Rest and motion are not "timeless" ideas, and representing them by a correspondence between ensembles of discretely distinct points is impossible. Quite surprisingly, there appears to be no mathematical axiom or theorem that holds that the real numbers (the arithmetic continuum) and the geometric line are the same thing.

Georgesçu argues that the arithmetization of physical phenomena rests on three axiomatic (that is, unproved/unprovable) claims. First, the "time continuum" and that of the reals are "identical." Second, the geometric line and the reals are identical. Finally, if a non-negative number is smaller than any positive number, then that number is zero. Given the usual assumptions on the structure of the reals (that is, well-ordering or some equivalent alternative), these axioms imply that phenomena such as evolution may be represented by arithmomorphic models. This basic faith is, to Georgesçu, quite misplaced. The arithmetic continuum was built up over time by the addition of classes of numbers composed entirely of discretely distinct elements. Thus, by adding together successive classes of numbers, each distinctly composed, we assert we have arrived at a continuous entity: the continuum of numbers.[9]

In order to illustrate a familiar set of "holes" in the arithmetic continuum, Georgesçu proposed the addition of a class of "infrafinite" numbers which he spent considerable time defining axiomatically.[10] Infrafinite numbers are suggested by a number of considerations in probability theory and analysis. For example, suppose x is a random variable, taking on any real values one wishes, with a left continuous cumulative density function $F(x)$. Then prob $(\underline{x} < x < \bar{x})$ for $\underline{x}$ and $\bar{x}$ real, $\underline{x} \neq \bar{x}$, is just $F(\bar{x}) - F(\underline{x}) \geq 0$. Suppose this difference is positive. Students of statistics are then understandably confused by the twin claims that, if F is smooth, prob $(\underline{x} < x < \bar{x}) > 0$, yet prob $(x = x^0) = 0$ for any x^0, $\underline{x} < x^0 < \bar{x}$. Clearly, though, the statement "prob $(x = x^0) = 0$" is not equivalent to the statement "x^0 cannot occur." If more advanced statisticians are not troubled by this, it is perhaps because they have gotten used to it. In any event, the same number (probability coefficient) is doing "double duty" for two quite distinct ideas: "prob $(x = x^0) = 0$" and "$x = x^0$ is impossible."

Another example of "infrafinite" numbers arises in plane geometry and is familiar to readers conversant with early speculative Greek mathematics. In particular, consider the "horn angle" type of curve in which two smooth curves converge at the horn's tip. Alternately, just consider a family of curves composed by, say, the functions $y = x^2$, $y = x^3$, and $y = x^4$ around the neighborhood of zero. Clearly, all these curves have derivatives of zero at $x = 0$, so the angles they make between the x axis and the curve are "the same" at $x = 0$. Yet one may find this unsatisfactory when one stares at the superimposed graphs. Here, infrafinite numbers allow one to distinguish between these angles in a natural way.

What do infrafinite numbers look like and how do they behave? Although a complete technical discussion would go well beyond the requirements of this book, a brief description is worthwhile. Cut the reals into two classes, the non-positives A and the positives C. Into this hole place the ordered set of all positive real numbers B. For such an infrafinite number $a \in$ B, we have 0 precedes a precedes r for any positive real $r \in$ C. We can accommodate $\alpha 2$ numbers by a lexicographic rule such that (r_1, a_1) precedes (r_2, a_2) if $r_1 < r_2$ or if $r_1 = r_2$, $a_1 < a_2$ for real components r and a. This is the lexicographic ordering of the plane. Further, $(r_1, a_1) + (r_2, a_2) = (r_1 + r_2, a_1 + a_2)$. Next, introduce a measure Meas (r, a) that satisfies Meas $(0, 0) <$ Meas $(0, a) <$ Meas $(r, 0)$ for any $a > 0$. Further, let Meas $(r, 0) = r$, implying Meas $(0, a) = 0$ for any a. Thus, any number $(0, a)$ is infinitely small with respect to any $(r, 0)$.

Georgesçu proceeds to develop a highly complex theory of these infrafinite numbers, relates them to the study of infinite series of functions by Paul de

Bois-Raymond, and argues that the notion of infrafinite numbers allows one to think sensibly about various "supertask" problems familiar in mathematical logic. Georgesçu's work is similar to that of Abraham Robinson (1974). For our purposes, the main points to keep in mind are: (a) the real numbers are discretely distinct entities; (b) real numbers are insufficient in number to represent some sensible orderings, such as horn angles; (c) "new" numbers are likely to be lexicographically related to the familiar "old" numbers. These conclusions are quite important for Georgesçu's epistemology and theory of science. In particular, models that represent some economic variables by real numbers will be called "arithmomorphic" and, due to the distinctly discrete nature of the real continuum, these models are inherently incapable of representing change. Further, sensible scientific propositions will have no analytic representation whenever these propositions are evolutionary in character. Yet these propositions, like the entropy law itself, may be valid, important scientific claims.[11]

Science and Arithmomorphism

Georgesçu's views on the structure of science, and particularly on the acceptable forms of scientific laws, are an important aspect of his economic philosophy and epistemology. Science, in contrast to many schools of philosophy, must in Georgesçu's view accept both being and becoming.[12] Thus, the correct representation of becoming is a central methodological issue in economics and the physical sciences. That economic processes involve change ("becoming") of a qualitative sort is central to Georgesçu's approach. The inability of analytic representations to accurately capture change, a consequence of the discretely distinct nature of the arithmetic continuum, implies that evolutionary laws will have a different form from those of mechanics. Economics, for historical reasons discussed in great detail elsewhere,[13] placed its methodological stock in mechanics with its time-invariant laws and quality-free structure. Yet the language of mechanics is inadequate to the discourse of economics.

Establishing the inability of mechanical paradigms to accurately model economic and social phenomena, which are evolutionary in character, is only part of Georgesçu's purpose, however. The more constructive aspect of his program identifies the forms of evolutionary laws, suggests that thermodynamics imposes an evolutionary character on economic processes, and offers a roadmap towards economic research that could be more relevant to the problems of survival of the human race. In order to explicate these features of Georgesçu's thought, we turn first to a brief description of his views on science.

Science, Georgesçu believed, is first a body of description or classification. Science is a "filing system." Yet, there are clearly very many ways in which facts can be sorted. Consideration of these problems of classification, including binary versus higher category classifications and self-reference, gave rise, in

Georgesçu's view, to logic and much of mathematics. Because there are very many ways to classify phenomena, one is naturally led to look for the "best" way.

Modern science has, over a very long period of time, developed the following basic structure. Propositions, which are (presumably) true statements about the interrelations of facts, may be classified into two types, A and B. All B propositions are implied by A propositions, while no A proposition is implied by any B propositions. While the B propositions are, strictly speaking, redundant, some are quite useful and are worth emphasis for their own sake.

New propositions are obtained through accidents, of course, although we can increase the probability of fortunate accidents through experimentation. Theory is used to suggest fruitful experiments.

"Theory", in its scientific sense, is ordinarily taken to mean a deductive, logical structure. Propositions of the B type are obtained from A type propositions using logical rules. While these rules do not, of course, guarantee that the A (and, hence, the B) propositions are *true*, this deductive process plays a vital role in science as it is usually conceived. This is most obvious in "sciences of essence," that is, mathematics, but it arises as well in the "sciences of fact" (for example, chemistry or physics).[14] Despite its role as an organizing principle of science, ordinary logic imposes a very strict limitation on the propositions upon which it acts. In particular, the use of logic in science requires discretely distinct meanings for both symbols and the concepts for which they stand. Thus, the concepts of science are taken to share a property of real numbers, "discrete distinctness." Logical operations are predicated on binary operators such as "is or is not," or "belongs to, or does not belong to." Models which attribute this property of the real numbers to concepts of interest to the analyst may also be called "arithmomorphic."

There are, of course, many concepts for which an attribution of discrete distinctness, or "utter separation from its other," seems meaningless. Examples of interest to science include numerous "qualitative" notions such as hardness, species, and so on, for which no satisfactory cardinal scales exist. One may very well include life itself in this category. For economists, the concepts of commodity, money, want, and many others do not exhibit a discretely distinct character. In Georgesçu's view, such concepts may be called "dialectical" in the sense that they are separated from the others by a "dialectical penumbra," rather than a distinct boundary which contains nothing. Thus, dialectical concepts violate the "principle of contradiction": both A and not A may be true.

The modern tendency among economists is to claim that dialectical concepts are "meaningless," which is equivalent to the claim that variables which cannot be represented by real numbers are meaningless. Since the concept of life itself is dialectical, one is then forced to conclude that the designation "living" must be meaningless. The positivist's objection is thus seen to be a reprise of the motto, "science is measurement."

Nor does it appear likely that one can dispense with dialectical concepts by claiming that they really reflect only imprecision in measurement. An examination of the concepts of, for example, "life" and "democracy" raise the question of what sort of technical progress will allow us to determine whether a society is democratic or a virus is alive. Science should serve man, not binary logic.

Although dialectical concepts are not discretely distinct, they are distinct. For example, although life may be a dialectical term, no one has successfully promoted the view that bricks are alive. Thus, although the concept "living" has a penumbra, that penumbra does not cover life's entire other.

It should be noted, and can be objected, that the penumbra itself is dialectical, else we would merely have three categories: A, not A, and "penumbra of A." However, the infinite regress implied by the penumbra concept is not an invalidation of that concept, according to Georgeşçu. Rather, infinite regression is seen by him as a necessary characteristic of any representation of change.[15]

Although arithmomorphic arguments are often successful in many applications, dialectical reasoning plays a crucial role in science. Perhaps the most important example, in Georgeşçu's analysis, is probability. Georgeşçu began his academic career as a statistician and he never lost his interest in the fundamental issues of probability, inference, and hypothesis testing. In common with several other commentators, Georgeşçu rejected both the subjectivist ("probability as degree of belief") and objectivist (Laplacean and frequentist) definitions of probability.[16] His rejection of what one might call the "naive frequentist" doctrine is instructive. Let fn be the relative frequency of some event E given n trials. Then the probability p of E may be defined as:

$$p = \lim_{n \to \infty} fn$$

Georgeşçu objects that a sequence of observations is itself merely a random event. Thus, he proposes an alternative "law of large numbers," for p, the physical probability of E, as follows. For any event E, there exists a number p such that, for any positive ε and δ, there exists an integer N so that $1 > \text{Prob} \, [|fn - p| < \varepsilon] > 1 - \delta$ for any $n > N$. It is instructive to note the difference between this formulation and the naive frequency doctrine described earlier. In particular, we observe here a point made by Georgeşçu in his critique of "mechanical" explanations for the entropy law. There is a difference between the claims that (a) one must wait a long time for a rare event to occur, and (b) if one waits long enough, a rare event will occur. This sort of confusion, which Georgeşçu argued was (is) endemic in statistics, in his view arose from muddled interpretations of the concept "random." In particular, he believed that "random" was a relation, not a cause, and that this relation arose because reality was dialectical and therefore could not, in principle, have an analytic representation.

Scientific Laws

The sciences of mechanics and geometry have a very special structure. In particular, the laws of mechanics and geometry describe relationships between cardinal variables that are timeless in either the sense that time plays no role at all or else that time is merely a cardinal magnitude that may be added or subtracted, and exhibits no preferential direction with respect to the phenomena being explained. Physicists describe this by saying that the equations of classical mechanics are "time symmetric." In general, cardinal magnitudes arise, in Georgesçu's view, from an idealization of physical operations such as "subsumption." The individual units composing a cardinal magnitude are completely devoid of qualitative variation. Classical physics concerns relations among magnitudes that are devoid of qualitative variation. Hence, order is unimportant and the laws are time symmetric.

That some branches of physics, such as classical mechanics, are unlike the bulk of science is not accidental. Physics is, in fact, the study of the relations between matter and energy that exhibit no qualitative variation. Contrast this with the science of metallurgy. Important concepts, such as "hardness," lack totally satisfactory cardinal measurements. (The Mohs scale of hardness is a completely relative one.) This is not a coincidence. Hardness, and a host of other similar concepts, are in Georgesçu's useful phrase, "quantified qualities." When one "quantifies" a quality, there is an irreducible qualitative residual remaining.

Economics, of course, makes fairly general use of "quantified qualities," as do all sciences except, perhaps, mechanics, several other areas of physics, and mathematics. What is the practical consequence of formulating laws using these "quantified qualities"? As Georgesçu was fond of noting, quantifying something or, more generally, representing something with a pencil and paper ("PAP") operation, does not change the nature of the phenomenon being quantified. If the variable of interest exhibits qualitative variation, then pretending that it does not is pretense only. However, while the representation of a variable does not affect the variable's nature, the converse is not true. In fact, quantified qualities appearing in "laws of science" leave a distinctive fingerprint: non-linearity.

In perhaps his most interesting and difficult metaphysical speculation, Georgesçu argues that the "natural laws of science" (which we should take in a broad sense) for cardinal variables must be proportionality laws only. Mechanical laws, for example, seem to generally satisfy this claim. Laws such as $e = mc^2$ have the form $e \propto m$; energy is proportional to mass, and so on. The component variables of these laws are cardinal and exhibit only variation in quantity. The linear forms of these laws are, according to Georgesçu, actually *required*: no true laws for cardinal variables can be non-proportionality claims. All natural laws among cardinal magnitudes must, in Georgesçu's view, be linear homogenous (that is, if $y = f(x)$ is linearly homogeneous, then f satisfies $\lambda y = f(\lambda x)$.)

The conjecture that all natural laws for cardinal magnitudes are linear homogeneous is, to say the least, a startling and perhaps quite original claim. One is reminded of Einstein's belief that a complete unified field theory ("theory of everything") would involve no "arbitrary" constants (such as the gravitation constant): the theory would explain all such values. More important for us than the truth of Georgesçu's idea is the source of this idea, for in this source we observe once again his intellectual preoccupation with the distinction between order and measure.

Although Georgesçu is not very clear on this point, his rationale for his conclusions on the forms of natural laws appears to be based on a type of "selection" principle similar in spirit to the Axiom of Choice. First, things that can be ordered must differ in quality, else ordering them would be impossible. One could not select elements to order unless some qualitative variation among elements made them distinct (though not, perhaps, discretely distinct).

Next, a natural law is a formulaic summary of a relation of some kind between two or more variables. One in effect selects "pairs" of values (x, y) and uses intuition or theory to infer a form of relation between the x and y variates. In the selection of these pairs, either the order of selection matters or it does not. In order for the order to matter, there must, however, be qualitative variation amongst the pairs. Such variation is ruled out, by definition, for any magnitudes (x, y) which are cardinal. Thus, for any law involving cardinal variates, total lack of qualitative variation implies that order cannot matter. Thus, their relation must be independent of the arbitrary cardinal scale and is therefore one of proportionality.

One is somewhat uncertain of what to make of this argument. Clearly, it exhibits an ontological sort of character, and relies on some propositions that are not very well specified. Nevertheless, the claim is interesting and is capable of casting a new light on some important economic phenomena. To see this, we first follow Georgesçu in noting that the converse hypothesis would presumably also be true: if the relation between variables is known to be non-linear, then these variables are quantified qualities rather than cardinal magnitudes.

The laws used in material sciences, for example, appear to offer a good demonstration of Georgesçu's idea. Almost none of these laws is linear, a consequence of the qualitative character of such concepts as hardness, elasticity, and so on. Consider the deformation of an iron bar as additional weights are pressed down upon it. As more and more weight is added, the rate of bending is not linear (the twentieth weight has a different effect than the first) and is not well approximated by a linear formula.

The sizes of living beings and life-bearing structures are relevant in this context. It is clear that organisms have optimal sizes and that life processes are not indifferent to scale. Georgesçu saw this proposition as essentially quite similar to the problem of the optimal scale of productive processes in economics. A common view among some industrial economists holds that, in the long run, decreasing returns to scale cannot occur. Such diseconomies are assumed to be ruled out by the possibility of exactly replicating some preexisting process and thus obtaining two parallel processes that result in exactly twice the outputs for exactly twice the inputs. Georgesçu termed such addition as "external": the constituent processes retain their individuality. Yet, it is clear that this is merely an exercise in amalgamation, and such amalgamation is completely unlike growth of an organism. We will have more to say on Georgesçu's model of production in the next chapter, although it is clear that he felt that "infinitely divisible production factors" and "infinitely divisible production processes" are not equivalent ideas.[17]

As should be clear, Georgesçu viewed all "real" change as necessarily qualitative. However, since qualitative change cannot in general be represented by analytical formulae, and since qualitative change is at the heart of life processes and the economic process, the question immediately arises as to both the feasibility of "laws of economics" and the nature of any such laws, if they exist. Georgesçu devoted considerable thought to this issue, and in several places made rather extensive comments on evolution and evolutionary laws.

He defined an evolutionary law as follows. Let E_t be some ordinally measurable variable at time t. Then an evolutionary law has the form: $E_{t1} < E_{t2}$ implies t_1 precedes t_2. We note immediately that laws of this type are not analytical formulae. Further, the index E_t need not be cardinally measurable, only "orderable." The Entropy Law, of course, serves as an admirable example of such a law. It is also clear (and methodologically important) that such laws can be discovered and articulated in the absence of a theoretical explanation for their existence.[18]

Although the evolutionary laws described by Georgesçu's definition are not analytical expressions, it is clear that, for such a definition to be relatively inclusive, ordinal measurability of some index E is required. Thus, we find that the notion of evolution is closely related to the notion of order. It is therefore not surprising that Georgesçu frequently argued that order is a more important category for economics than is measure. One can certainly imagine change which is not evolutionary, and we might call such a process change *per se*. For example, cycling is change but is not evolutionary. More technically, one could imagine a sequence of randomly selected real numbers. Neither of these examples exhibits the feature of order which defines evolutionary changes.

History, of course, is an evolutionary process characterized, according to Georgesçu, by hysteresis and novelty. Despite this, and over the probable objections of committed Marxists, Georgesçu felt that we have "found no evolutionary laws of society." An examination of the conditions specified in the Entropy Law can illustrate a possible cause of this failure. An evolutionary process takes place in a certain environment. For human societies, the nature of this environment is not stable, but involves gross intrusions such as wars with neighboring societies, natural disasters, and the like. Under these conditions, discovery and/or identification of an evolutionary law for a human society is difficult.[19]

The concept of an evolutionary law is inextricably bound up with the notion of Time. That Time has served as the fountainhead for many paradoxes over millennia is, for Georgesçu, the natural consequence of man's attempts to represent Time arithmomorphically. Yet Time, and our direct experience of Time, is dialectical, seemless, and overlapping. No representation of Time is possible using the arithmetic continuum, and attempts to do so produce enigmas such as Zeno's paradox. One may object that "mechanical time," such as is represented by the movement of hands on a clock, appears to serve science quite well in many applications. This is true, yet in Georgesçu's view scientific laws that refer to clock time (denoted t) differ fundamentally from laws, such as the Second Law of Thermodynamics, that refer to the flow of Time (big T) *through an observer's consciousness*. Time (T) causes the clock to move (t), not the converse. Laws that refer to t, such as the mechanical laws of pendular motion, are not causal claims, but instead statements of *coincidence*: one mechanical system (for example, a pendulum) is merely in phase with another (a clock). Time with the capital T, however, is a cause, in a specific sense, of changes summarized by indices used in evolutionary laws.

In the *Entropy Law*, Georgesçu devotes considerable discussion to the view associated with, among others, the philosopher J. McTaggart,[20] that "time" is unreal. In a somewhat abusive simplification, McTaggart's position can be summarized as follows. First, the order in which events occur, that is, the "earlier than" relation, is itself timeless: the Battle of Hastings is earlier than the Battle of the Bulge, and has "always" been so and always will be. Our notion of time arises, in fact, from "memory." Time, according to McTaggart, may be supposed to be unreal and no contradiction arises.

McTaggart's position is not an uncommon one among certain philosophers of the twentieth century. However, based on our previous discussion, it is obvious that Georgesçu could not accept such a conclusion. In particular, McTaggart's position, at least as we have summarized it here, appears to identify our experience of Time (capital T) with time (little *t*) as represented by the arithmetic continuum.

This discussion also highlights a central tenet of Georgesçu's methodology: evolutionary laws must be stated in terms of T and not *t*, and such laws cannot have analytical expressions. For example, the Entropy Law states that the entropy of an isolated system cannot decrease as Time passes, but this law does not claim how quickly entropy will accumulate. This is not a coincidence. Statements of the latter kind, amendable to analytical representations, can never, in principle, articulate evolutionary laws.

3.2 ECONOMICS

Economic Methodology

Having reviewed, if only in a very superficial way, Nicholas Georgesçu-Roegen's main epistemological assumptions, we turn now to a discussion of economic methodology. Here again, Georgesçu's writings provide a fairly rich, if unorganized, body of work that can be used to distill his main methodological preoccupations.

Economics, in Georgesçu's view, is not, and cannot be, a "theoretical science," that is, a filing system for facts in the "axiom" and "derived theorems" categories. The boundaries of such a science are entirely too narrow to contain the purposes of economics. In contrast to the largely unexamined intellectual underpinnings of neoclassical economics, Georgesçu felt that a primary purpose of economics was *not* to facilitate positivist calculations superficially devoid of value judgment, but rather to provide a framework within which we could discuss the "best aims for Mankind." Thus, rather than arguing that economics provides a way to calculate the "best means to a *given* end," Georgesçu was interested primarily in the *ends*. Although not explicitly stated in most cases, he regarded the long-term survival and maintenance of the human race as the primary end, and much of his work on ecological economics, resource scarcity, and similar matters can be properly regarded as an effort to *change people's minds*. For example, Georgesçu rejected the idea that problems of integenerational resource allocation were primarily conventional problems of "finding the right prices." Prices, after all, are the effects of preferences, not the causes of preferences. Unless people's opinions are changed, even the implementation of a price/tax scheme to slow resource depletion is problematic.

Some insight into Georgesçu's views on the proper institutional assumptions for economic modeling is obtained from his well-known critique of neoclassical, "mainstream" economics. In simple terms, Georgesçu viewed neoclassical economics as being composed of: (a) wealth maximization hypotheses; (b) positive preferences by agents for "more"; (c) diminishing marginal utility assumptions; (d) decreasing returns assumptions; and (e) an assumption of the disutility of labor. Among the practical consequences of this framework are marginal pricing rules for the production of goods and employment of factors.[21]

Although not immediately obvious, it was clear to Georgeşcu that the above characterization of economic life is not "institution free." In particular, he argued forcefully that the institution enshrined in these hypotheses is the bourgeois life of the towns. No economic model is "institution-free," and certainly town life is not typical of the circumstances of much of humanity. In particular, as we will observe in the next chapter, neoclassical analyses and marginal pricing rules may be inconsistent with the institutions of, for example, peasant economies.

Georgeşcu-Roegen firmly believed that economic models needed (indeed, required) institutional detail. Since institutions vary among cultures, no single theory can be expected to explain every economic phenomenon. Although arguments of this sort are familiar, the deeper significance of Georgeşcu's position could easily be overlooked. An unspoken assumption, used by almost all mainstream economists, is that there is some true "model" of economic behavior, of which various societies and markets represent different "parameterizations." For example, it is often implicitly assumed that economic development proceeds as a process of capital accumulation and technological change that, in a broad sense, applies to all societies. The economic experiences of the world's peoples are thought to differ not because some are utility maximizers while others are not, but because the "initial" conditions of societies vary widely. Of course, if "initial" conditions are taken to include everything, such a view could hardly be wrong, nor would it be useful. There is, however, a historic tendency to err in the other direction by, in Georgeşcu's view, attempting to explain every economic phenomenon in the world using profit maximization or some similar tenet.

Yet it would be quite incorrect to lump Georgeşcu among institutionalists such as Ayers. Georgeşcu was quite willing to use the neoclassical analytic arsenal when he felt it was justified by the problem at hand. Ironically, his criticisms of analytics were of an essentially mathematical character. More importantly, his views on the importance of institutions and culture in economic modeling arise from an early recognition of the relationship between certain phenomena and "scale." In particular, there are many characteristics of human societies that do not, and cannot, occur on the scale of individual agents. Just as there is nothing in the natures of either hydrogen or oxygen that lead one to be able to predict the curious physical properties of water, there is likewise nothing in the view that people are utility maximizers that explains the nature of the Catholic liturgy, for example. "Institutions" is simply the broad label assigned to such phenomena. Although it is common in some neoclassical traditions to "explain" institutions by referring to, for example, the profit motives of various agents, such a course is akin to explaining the properties of water by noting the properties of its elemental constituents. Ultimately, atomism of this kind may not be "wrong," but it does rest on a powerful and unarticulated assumption: all properties of combinations can be analytically explained by the simple properties of the constituent elements. That this assumption is incorrect in the physical world is clear.

Georgesçu viewed the evolution of social institutions as mildly similar to evolutionary processes in the natural world. "Mutations" (that is, novel institutions) arise from "accidents" and man's love of variety and novelty for their own sakes. Traditions or cultural norms play the role of those physical processes which apply brakes to the promulgation of genetic mutations in species. In particular, most innovations are bad and, even for the good ones, the passage of sufficient time is required to determine if the innovation contributes to the fitness of the society/species. Thus, traditions and other similar social rigidities, rather than representing either a contemptible reaction to progress (as Marxists would claim) or a carefully crafted bundle of contracts between strategically aware, self-interested agents (as some neoclassicals would believe), do serve an important practical purpose. Most importantly, this purpose is not the result of conscious contracts by individual agents. The emergence and evolution of such institutions as peasant village land usage conventions is a process that occurs at the *social* level of organization, and must, Georgesçu felt, be analyzed in that light.

Economic Models

Georgesçu's views on epistemological and methodological matters, if accepted, have a number of practical consequences for the creation of economic models. Perhaps the most important of these consequences for contemporary economic practice is the notion, essential to Georgesçu, that the economic process is an evolutionary one, involving irreversible change in one direction only, that cannot be represented by a mechanical analogy. One useful way to restate this idea is to argue that Georgesçu saw the economic process as one in which either (a) finite stocks of things are exhausted through time or (b) finite stocks of other things are accumulated through time. Entropy, of course, provides an example of something that accumulates, while minerals in exploitable concentrations offer an example of something that is exhausted.

Although the inclusion of such stocks in models of economic activity over time may seem merely a small innovation, this innovation completely changes the nature of such models, destroying the "steady states" which were the models' most prominent prediction.[22] Growth (as opposed to development,

à la Schumpeter) cannot continue indefinitely. Georgesçu formalized this point
as follows. Let S represent the finite stock of some accessible resource (or of
all resources). Allowing Pi to equal the population at time i, and Si the part of
S depleted per person at time i, then the maximum quantity of human life
$L = \Sigma Pi$ is subject to the constraint $\Sigma PiSi < S$, so that $Pi = 0$ for all $i >$ some future
date n. The "duration" of the human species, n, is also presumably finite.[23]

Viewing the economic process as an irreversible, unidirectional one also has
important implications for models of human welfare when certain facts about
the order in which resources are exploited are recognized. Georgesçu notes
repeatedly in his writings that resources such as mineral deposits are, on average,
exploited sequentially in order of concentration, with better grades being mined
first.[24] This observation appears to be quite general and arises naturally from
the finiteness of the stocks *in situ* and the effects of technical progress.

As the desired material is extracted and purified from the ore, waste material,
which is often toxic, is generated. The minimum amount of such material
obtained, given an average concentration of ore, is proportional to the quantity
of refined ore obtained. As ore concentrations fall, larger and larger volumes
of waste are generated. Due to the First Law of Thermodynamics, no technological advance can, even in principle, alter this relationship.[25]

The analysis above illustrates, if only in a very simple way, a few of the
practical implications of recognizing the role of accumulating or decumulat-
ing stocks of various kinds in economic models. Of course, many neoclassical
economists, such as the late Julian Simon, would argue that *ideas*, not natural
resources, are both the motive engine of human progress and can be available
in "unlimited" quantities. This view, which is quite familiar in western
economics, might be termed "technological optimism." That Georgesçu did
not share this optimism is well known. The sources of his pessimism should now
be fairly clear.

Georgesçu's research into the Entropy Law, and his proposal of a "Fourth
Law" of thermodynamics (a closed system, receiving only energy from outside,
cannot perform work at a constant rate forever: matter matters) led him to adopt
the very old economic idea, associated with the Physiocrats and still used by
many Austrians that some industries or sectors are "prior" to others.[26] This
notion – that industries can be, in some meaningful sense, "ordered" in a vertical
fashion – is entirely absent from standard economic models. In Arrow–Debreu
or Leontief representations of the economy, ordinarily the output of all sectors
may serve as inputs in all sectors. Since "everything depends on everything
else," it is not meaningful to speak of one industry as "parasitic" on another.
Georgesçu's understanding of the relevance of thermodynamics to the
economics of production enabled him to realize that "low" entropy was essential
to any activity we call economic. Industries such as mining provide this
irreducible requirement and can, in entropic terms, be regarded as "prior" to

manufacturing activities. This is not to say, of course, that the outputs of other sectors are not used by the mining industry. Rather, these other outputs are themselves dependent on mining outputs in a way mining outputs are not dependent on them. We could produce ores or iron without automobiles but the converse is not true. This sort of vertical dependence arises because mining amounts merely to gathering of "low entropy matter" from nature's endowment, and it is this endowment that makes other activities possible.

Closely tied to the special status Georgesçu accorded natural resources is his view of technical change and growth, which are intertwined with his ideas on dialectics, dimensionality, and the like. Call the set of technical recipes available to humans at some time "technology." Many economic operations, such as mining the moon, could be called "feasible" in the sense that operations that would affect this result are known. Such a course, however, could not be called "viable," in the sense that such mining consumes more resources than it can produce. Lunar mining cannot be the basis of any ongoing social system since the proceeds of the activity are insufficient to support it. Thus, a "viable" technology is, subject to the limited stock of resources on which it is based, capable of "reproducing itself." Coal-fired steam engines are viable in this sense, because by use of such an engine, mining coal and melting ores in sufficient quantities to produce many more steam engines is possible. This expansion can continue until accessible coal stocks are exhausted.

Given any existing technology, the emergence of a new technology depends on both a novel insight and the ability of the existing technology to fabricate the trappings of the new. As Georgesçu put it: "The first bronze hammer, for an unadorned example, was produced by stone hammers." Thus, a viable technology "must be capable of reproducing itself after being set up by the technology now in use."[27]

This view of technology has implications for issues of growth and economic development, and the distinction between them. Growth – the expansion of the rate of operation of existing production processes – is seen as a very confusing idea. How can it occur? One cannot say there is an "increase in resources" in the sense of increased outputs of primary materials, since this could not happen without an increase in those inputs that these sectors require. That, in turn, requires additional increases in other inputs, and so on. A flow cannot contemporaneously fund its own increase. In the Leontief model with a single primary factor, labor, growth (in the absence of technical change) must require an increase in the length of the working day and a decrease in the output of leisure or a population increase. Indeed, Georgesçu attributed a good portion of the modern West's increased economic output to just such a lengthening of the hours of work.[28]

Development, on the other hand, requires a new idea that leads to a viable technology. Such changes may, of course, make possible growth in its arithmetic sense as well. Georgesçu consistently advocated development over any narrow focus on growth in its purely quantitative sense, although the latter is the focus of neoclassical research.

Consideration of Georgesçu's views on growth and development illustrate a final and quite important aspect of his thought. Although Georgesçu-Roegen was a highly trained mathematician whose teachers included Fréchet, Pearson, and many others, he was, first and foremost, a practical man. To him, variables used in economic models were supposed to represent something real, even tangible. This simple requirement is seen, in retrospect, as the basis for many of his criticisms of neoclassical and Marxist economics, particularly in the areas of utility theory and production models. Insisting that the magnitudes underlying the variables be specified results in several surprising consequences. First, the dimensionality of all terms must be specified. Second, the logical requirements for growth or change to occur are made more explicit. Third, economists are cured of the disease, which is pandemic, of attributing properties associated with variables used to *represent* states with the states themselves. In Georgesçu's telling phrase, "pencil and paper" operations do not cause real changes in anything. This preoccupation with the exact meanings of variables, their dimensions, and their interpretations can be seen at the heart of many of Georgesçu's contributions to economics.

The Bioeconomic Program

The culmination of Georgesçu-Roegen's methodological concerns is the "Bioeconomic Program," an ambitious, though incompletely articulated, research proposal fusing economics, physics, biology, sociology, and political science into a coherent whole. The unifying theme of this program is the view that mankind's purposeful behavior both influences and is influenced by the natural environment. Economics itself is sometimes termed by Georgesçu the "study of changes in matter and energy produced by human actions and entropy." Thus, economics is seen as a manifestation of man's biological struggle for existence. As bees build hives, collect nectar, and so on, Man builds houses, farms, and engages in manufacturing. For both men and bees, a natural environment which provides usable energy and matter of suitable types and availabilities is a necessary and indispensable requirement for survival. By understanding these processes, economists should seek to promote the long-term continuation of the human race.

The view that economics is an aspect of man's biological existence is not original to Georgesçu. Early analysts, such as the Physiocrats and Malthus, recognized such linkages even though they did not articulate them in modern terms. Perhaps the most important precursor in this regard is Marshall, whose "organic" view of industrial evolution, and whose writings touching on biology, were frequently praised by Georgesçu.[29] Nevertheless, Georgesçu's conception of the connection between the economy, environment, and human society is unique and uniquely articulated. Unlike earlier writers, he had a mechanism – thermodynamics – which provided an explicit, though not analytic, path for interaction. Human action accelerates the entropic degradation of the environment. Manufacturing results in a transformation of available energy and propitiously concentrated matter into bound energy and dissipated matter. The history of agriculture is one of increasing substitution of finite, "low" entropy mineral resources for relatively abundant solar power harnessed through organic farming practices, draft animals, and the like. Because the consequences of such activities cannot be reversed, human actions today influence the quantity of future human life supportable by the earth.

Georgesçu devoted considerable thought to the problem of the intergenerational allocation of resources, and held the opinion that markets for these resources led to "excessive" exploitation of them. Although we will examine this issue more carefully in a later chapter, the basic intuition of his claim, and its probable defects, is easily given. First, "correct" prices for resources arise through bidding by all interested parties. If some relevant bidders are left out, prices are necessarily either correct or too low: prices are non-decreasing in the number of bidders. Second, due to entropy, use of resources by current generations precludes their use in an equivalent way by future generations. The economic process transforms "low" entropy resources into a "flux of enjoyment" for current users, and higher entropy outputs of waste. Thus, resource exploitation decisions irrevocably increase the stock of higher entropy materials left to future generations and reduce the stocks of desirable materials accessible to any given technology.

Finally, future generations cannot bid on current resources and, unless the preferences of current owners change, the absence of future demands from current markets will result in prices that encourage excessive exploitation.

Although this argument is plausible, it is apparent that numerous complicating factors must be dealt with in order to assess its validity. First, as Stiglitz points out: "one should not view equity in the narrow sense of simply looking at the division of natural resources between present and future generations; the present generation may give future generations fewer natural resources . . . but it will give future generations a higher level of technology and more capital."[30] Additionally, we must face the issues of what it means for resources to be used up "too fast," and the roles played by overlapping generations and markets for natural resources in inducing current consumers and producers to act as "agents for posterity."[31] Substitution elasticities also clearly matter. We will deal with these issues in greater detail later in the book, but it is important here to emphasize the paramount role Georgesçu attached to the issue of intergenerational equity.

Conventionally trained economists, surveying the argument outlined above, will almost surely point out the role of technical progress in increasing human welfare. Indeed, one can quite plausibly argue that, though the world of the latter fraction of the twentieth century has, for example, far fewer mineral resources than the world of Christ's time, nevertheless living standards are incomparably higher for most people. This has occurred because technical advances make it possible to harvest previously unusable forms of terrestrial "low" entropy, such as petroleum and uranium.

Georgesçu did not share the technological optimism of many economists, and the reasons for this pessimism are rooted in his methodological outlook. First, the economic process resides in historical time. The circumstances of the last 100 years, often termed a period of "mineralogical bonanza" by Georgesçu, may not be representative, and one cannot extrapolate to discover the course of an *evolutionary* process. More prosaically, the input "natural resources" is incorrectly treated by neoclassical economics as being symmetrical with other inputs such as capital or labor. If one plans to substitute capital for mineralogical resources, Georgesçu insists that we ask: Of what will the capital be made? Is it possible, as a simple Cobb–Douglas model would suggest, to substitute other inputs for resources to any required degree, so that natural resource input flows can be made arbitrarily small? Can automobiles, say, be made of capital and labor only?

Georgesçu saw much of the technical optimism of conventional economics as arising from two primary fallacies. First, the use of mechanical analogies has led economists (and others) to think of the economic process as itself mechanical, thus obscuring the importance of entropic degradation and the exhaustion of terrestrial stocks of natural resources for economic reasoning. Mechanical, arithmomorphic models are incapable of representing the ecological problem correctly. The Entropy Law imposes a constraint on economic activity that is not analytic and is unlike the constraints on optimization problems familiar to economists.

A second error, related to the first, concerns the assignment of properties relevant only to analytic representations of production to actual production processes. Pencil and paper operations do not affect reality, and the mere fact that a mathematical production function suggests substantial substitution possibilities does not mean such potential actually exists. It is the idea of unlimited substitution that is incorrect and our mathematical treatment of substitution cannot alter this incorrectness. As we will see in Chapter 5, the Entropy Law imposes an energy requirement on materials processing that can never be avoided.

A final error is made when one fails to recognize that technical progress, the "savior" of neoclassical growth predictions, is a dialectical, not an arithmomorphic, concept. The irony in this is apparent.

The bioeconomic program remains incomplete. Although Georgesçu wrote extensively on bioeconomic problems, his advertised book on the subject was never published.[32] It is also unclear that thermodynamics provides a unifying template of a power comparable to the optimization/equilibrium scheme of conventional economics. One sees in bioeconomics, however, the unique result of a unique vision of economics and science. Yet bioeconomics did not spring fully born from Georgesçu's pen, but developed very slowly over the course of his long career. The first part of that career focused on more conventional analyses of issues we might term "normal science," though many of the methodological issues discussed in this chapter emerged first in connection with mainstream projects. That work is the subject of the next chapter.

NOTES

1. See Chapter 2 for a detailed biography of Georgesçu-Roegen.
2. Georgesçu stated that his interest in "non-standard" problems arose from his dissatisfaction with neoclassical models applied to agrarian societies such as Romania. He noted: "It was evident to me that standard economics could not represent an agrarian economy . . . I thus acquired a special eye for issues ignored by the standard economic persuasion or by ordinary economic analysis" (in Szenberg, 1992: 129).
3. This idea first occurred to Georgesçu in the mid-1930s. It remained novel enough so that Samuelson commented on it in his contribution to Georgesçu's retirement volume *Evolution, Welfare, and Time in Economics* (Tang *et al.*, 1976).
4. The familiar Pythagorean legend holds that the disciple who proved that the length of the diagonal of a square with unit sides was not commensurable was taken to sea and drowned by enraged colleagues.
5. Cantor led a rather tragic life, suffered from depression, and spent his academic career at lesser universities. See Boyer (1985).
6. "Aleph" is the first letter of the Hebrew alphabet. The ancient Greeks likewise used an alphabetical numbering nomenclature.
7. The interested reader is urged to study Appendix 1 of N. Georgesçu-Roegen, *The Entropy Law and the Economic Process*, 2nd edn, for an extended discussion of these issues.
8. This view was, of course, forcefully argued by Bertrand Russell, who further held that such a conceptualization was completely adequate. See Russell (1903).
9. This is not to say that the reals R were literally defined as the union of these "earlier" sets.
10. Georgesçu-Roegen, *The Entropy Law and the Economic Process*, Appendix 1.
11. The reader familiar with Henri Bergson will note the resemblance between some of Georgesçu's propositions and those of Bergson. Oddly, Georgesçu's criticisms of the over-mathematicalization of economics are primarily of a mathematical character.
12. Georgesçu-Roegen, *The Entropy Law and The Economic Process*, ch. III.
13. A fascinating and controversial history of physics influencing economics is Mirowski (1993).
14. The distinction between "sciences of fact" and "sciences of essence" was very important to Georgesçu, but is "dialectical" only. This is treated below.
15. It is unclear whether Georgesçu implied that this "infinite regression" phenomenon is inherent in the arithmomorphic character of the *representations* of change or inherent in change itself.
16. See "The nature of expectation and uncertainty", in Bowman (1958).

17. Any phenomenon that is not indifferent to scale is represented by "quantified qualities" and their non-linear laws. See Georgesçu-Roegen, *The Entropy Law*, chs 4 and 9.
18. Indeed, though economists seem to believe that physical laws are derived mathematically, this is not always true. The Entropy Law is an empirical generalization. Deriving it from statistical arguments is a process of which Georgesçu was a vehement critic. See ibid., chs 5 and 6.
19. An interesting possible exception would be a linguistic law that asserted, for example, that over time consonants "soften" or otherwise mutate in some predictable manner.
20. McTaggart (1927). Of course, this idea (that time is unreal) is very old, and is particularly associated with St. Augustine as well as with Platonists of all stripes.
21. That such analyses could not be applied to agrarian economics was, in Georgesçu's view, an important indictment of mainstream economics. However, in fairness to neoclassical economics, the outline offered by Georgesçu is no longer the only paradigm nor, perhaps, the dominant macroeconomic story. The role of information, in particular, is missing.
22. Toman *et al.* (1995) offer a very useful discussion of neoclassical growth models and resource/environmental depletion.
23. Complicating all this is the role of technical change and the like. Toman *et al.* (ibid.) give a good discussion.
24. An extensive, if somewhat unsatisfactory, discussion of this appears in Georgesçu-Roegen, "Energy and economic myths" (1975).
25. Admittedly, the environmental consequences of the relationship can be affected by technical advances in waste storage and the like.
26. Students of the Austrian School will recognize the notion of "roundaboutness" here.
27. Georgesçu-Roegen, "Technology assessment: the case of the direct use of solar energy", *Atlantic Economic Journal*, **6** (4), 1970: 18.
28. *The Entropy Law* contains many suggestions along these lines.
29. Some of these ideas appear in Marshall's *Principles of Economics*, where Marshall acknowledges that neither energy nor matter can be created by humans.
30. Stiglitz, in Smith (1979: 61).
31. The reader may recall the *bons mots*, attributed to Boulding: "What has posterity done for me?"
32. Princeton University Press was listed by Georgesçu as publisher of the volume.

4. Georgesçu-Roegen and "normal science"

4.1 INTRODUCTION

Georgesçu's most controversial writings, such as those on the Fourth Law of Thermodynamics, intergenerational resource allocation, and public policies to promote environmental ends, are quite controversial indeed. As a result, many economists, particularly those under 50 years of age, are familiar with Georgesçu solely in connection with his environmental advocacy. Yet he had a long and quite distinguished career in conventional economics, particularly in mathematical economics. His vast output of technical work is, by itself, more than sufficient to justify his inclusion among important modern economists.

Aside from some of Georgesçu's very earliest work in utility theory and his analysis of substitution in Leontief-type models, virtually all his contributions to what Kuhn termed "normal science" are not particularly "normal." Rather, time and again, Georgesçu uses an analysis of a "conventional" problem to launch a novel program of research. Often, the impetus for his alternative theories of, for example, choice and production, lies in a deep, probing, and thoroughly practically minded examination of the underlying formal structure of the conventional model. For Georgesçu, economic models are analytic "similies" that, nevertheless, utilize terms or variables which must correspond to some real coordinate of a physical process. Given this attitude, issues that rarely trouble "conventional" analysts troubled Nicholas Georgesçu-Roegen a great deal. As a consequence, themes such as hysteresis, evolution, the representation of important economic coordinates by ordinally measurable variables (that is, real numbers), and the dimensional homogeneity (comparable units) of variables, naturally emerge in Georgesçu's work. His approach to economic theorizing, in fact, almost guarantees that these themes, which seem quite byzantine to conventionally trained economists, must come to the fore.

Because Georgesçu's contributions to conventional economic analysis are so extensive, it is difficult to select a reasonable number of topics for inclusion. Any selection is likely to omit papers of potentially great interest to some readers. Thus, some criteria must be applied. In selecting the topics to be covered here, the following factors were considered. First, papers widely cited in texts or journal articles are usually included. Thus "The pure theory of consumer's behavior" and Georgesçu's contributions to the Leontief model are present. Second, papers which, in some epistemological sense, form a related body of work may be selected as "jointly important": Georgesçu's work on consumer hysteresis and his hierarchical model of wants qualify under this criterion. Third, topics of great human importance, such as capitalist breakdown or agrarian policy, are included. Finally, topics which Georgesçu himself indicated were of importance, either through casual conversations with colleagues and students or the selection of work for reprinting in scholarly retrospectives, can be included.

4.2   GEORGESÇU AND UTILITY THEORY

Nicholas Georgesçu-Roegen's earliest published contributions to formal economics examine utility theory, a topic that interested him throughout his life. His Harvard experiences focused his attention on several classic problems in the theory of consumers, including the so-called "integrability problem," and the issue of welfare evaluation and the "constancy of the marginal utility of income." Georgesçu made important, indeed fundamental, contributions to these and a few other issues in the mid to late 1930s. It is to some degree a testament to the technical brilliance of Georgesçu that these papers continued to be actively cited as relevant (as opposed to merely historical) references well into the 1970s.[1]

Georgesçu's work on consumer behavior is important for at least three reasons. First, the work itself was important and has been incorporated into graduate texts in economics. Later commentators, such as Samuelson, Chipman, and others, have recast Georgesçu's analysis into more modern forms that take advantage of duality results, though this luxury was unavailable to Georgesçu at the time. Second, many of the themes with which he became associated in later years, such as hysteresis, imperfections of measurement, ordinalist fallacies, and the like, were brought to his attention by his introspective examination of consumer choice. As Georgesçu frequently argued, consumer choice is a topic which, unlike problems in, say, physics, is quite legitimately a subject for "mental experiments," that is, introspection. This refreshing lack of formalism remained with him throughout his life. Finally, Georgesçu obtained a very good reputation as a mathematical economist largely through his work on consumer theory, and secured the informal patronage of Schumpeter, Samuelson, and others as a result. The "neoclassical credentials" represented by his work on utility theory (and, later, input–output models) undoubtedly secured a somewhat less hostile reception for his subsequent, heterodox work on thermodynamics and methodology in economics. It was more difficult to dismiss as a crank an economist who had made fundamental, neoclassical contributions, equations and all. Thus, unlike so many critics of mainstream economics, such as John Kenneth Galbraith or Kenneth Boulding, Georgesçu enjoyed a significant reputation in the world he was at pains to overturn, the world of highly technical mathematical economics.

Georgesçu's approach to choice theory, especially in his earliest writings, is quite likely to strike the modern student as peculiar. In particular, he frequently refers to "stability" of consumer's optimum (which we would now just call strict convexity), and makes numerous statements of the form: "the consumer thus finds it worthwhile to take the trouble to move in a preferential direction", or something similar.[2] As was the case with many discussions of choice theory in the 1930s, the complete separation of consumer behavior into preference components and the acts of choice was not yet complete.[3] Of course, this distinction would never be complete or even regarded as meaningful by Nicholas Georgesçu-Roegen, as we shall see below.

Georgesçu made many contributions to choice/utility theory, and it is impossible here to explain each in detail. Nevertheless, several ideas deserve special treatment because they are either: (a) historically important; or (b) novel insights still largely unexplored in economics. First, we present a brief overview of the "integrability" problem and Georgesçu's contribution to it. Next, we examine "threshold" phenomena (stochastic choice models), hysteresis in preference, hierarchies of wants, and critiques of the axioms of expected utility theory.

Georgesçu and the Integrability Problem

The "integrability problem" refers to several issues in the relationship between preferences and market-observable phenomena such as demand curves. In particular, the integrability problem is often stated in modern economics texts in a form similar to that offered by Varian (1992):

> Suppose that we were given a system of demand functions which had a symmetric negative semidefinite substitution matrix. Is there necessarily a utility function from which the demand functions can be derived? This question is known as the integrability problem.[4]

As described by Varian, the integrability problem is actually a complex of several closely related issues. First is an existence problem: do there exist preferences that "rationalize" some demand system? Second, what properties of demand systems are mirrored, and how are they mirrored, in the corresponding preference field? Third, how would one, in *principle*, find the preference field from the demands? Finally, how in *practice* would the preferences be found? This division of the problem into related parts is not explicit in most accounts of this debate from the 1930s.

Before describing the problem as understood and solved by Georgesçu, it is useful to briefly review the integrability issue as it now stands.

Given his preference field, a consumer selects a "most preferred" bundle of commodities subject to his budget constraint. Repeating this choice, demand curves are obtained, which are behavioral curves with various properties inherited from preferences and the maximization process. "Global" properties of these demand curves are described by the weak and strong axioms of revealed preference,[5] denoted here by WARP and SARP, respectively. "Local" properties are taken ordinarily to be derivative properties of these curves, and such properties are summarized by negative definiteness *and* symmetry of the corresponding second derivative "substitution" matrix, in either prices (Antonelli) or goods (Slutsky) forms. The WARP yields negative definiteness of the corresponding substitution matrix (cross-price derivatives) but not symmetry unless there are only two commodities. Thus the WARP is insufficient to establish existence of a "utility function" that yields the subject demands (Gale (1960)). On the other hand, the SARP is sufficient, as shown by Houthakker (1950) in his famous paper.

The "integrability" nomenclature arises because, to "recover" the preferences from the demand system, one "integrates" a system of differential equations which contain demands as arguments. The mathematical requirements for this system to have a solution turn out to be symmetry conditions on the substitution matrix. The intuition for this is that, under symmetry, a single integral surface passes through each point. Yet – and this was not understood in Georgesçu's early days – the mere fact that one can "integrate" up to the "integral varieties" does not guarantee the finding of, nor even the existence of, an underlying utility function ("ophelimity" function in Georgesçu's usage, courtesy of Pareto) that generates the demands. This is because integrability in the mathematical sense is insufficient: the demands are supposed to "maximize preferences." Economic integrability requires that the substitution matrix be negative semi-definite, so that the "integral function" obtained is maximized by the demands. Thus, the "integral varieties" obtained by solving the system of partial differential equations (or its corresponding total differential) are not the indifference curves ("indifference varieties" in Georgesçu's usage) unless these additional conditions obtain. Put simply, while mathematics requires a family of non-intersecting integral surfaces, economics requires that these surfaces be convex. This additional restriction is not required by the Frobenius theorem, which is the primary result needed to establish integrability in its mathematical sense.

Modern treatments of the integrability problem make use of a number of theoretical advances unavailable at the time of Georgesçu's early work on this problem. Samuelson later showed how one passes between symmetry of the price substitution (Antonelli) matrix and quantity substitution (direct or Slutsky) matrix by inverting. Further, integrability can now be obtained even when one representation of demand, for example, the direct demands $Q = f(p)$ (where Q is goods demanded and p is the price system), is not single-valued, while the alternative representation $p = g(Q)$ is. These generalizations allow for integrability even when the indifference curves in quantity space have flat portions (so $Q = f(p)$ is not single-valued) or kinks (so $p = g(Q)$ is not single-valued). Finally, duality theory now offers economists a large set of equivalent representations of the same problems so that, given the purpose at hand, easier systems can be studied. As a consequence, the presentation of the integrability problem in most modern economics texts looks very different from that studied by Georgesçu.

When Georgesçu began his work on integrability, he started at a theoretically important point reached by Pareto, who, however, did not really know what he was doing. In particular, Pareto proposed two notions of how one could pass from market data (demands) to preferences. The first, Georgesçu's starting point, is based on the following template. Select some price system p. Associated with it is an optimal consumption bundle $[X^*(p)]$. Now, vary p and see how $[X^*(p)]$ varies. By doing so one has generated a map of the marginal rates of substitution at all bundles. This is the intuition behind Georgesçu's approach.

More formally, Georgesçu begins his first important paper, "The pure theory of consumer's behavior" (1936), in the proper way by introducing four axioms or postulates and a related set of elements. These axioms, labeled assumptions by Georgesçu, are quite modern in spirit, if stated in a somewhat old-fashioned, verbal form. Assumption A, which, it turns out, represents an idea that was to have tremendous significance for Georgesçu's writings and career, dispenses with the "ordinalist's fallacy," an error rampant in all choice theory models of the day.[6] Imagine a "preferential set" of bundles (a set of bundles constituting, for example, a positively sloped line in the commodity plane) which can be placed in one-to-one correspondence with the real numbers in such a manner that those bundles assigned a higher number are preferred. Then, if there are two bundles in the set C_α and C_β such that C_α is preferred to some fixed bundle C_T, and C_T is preferred to C_β, then there must exist a bundle in the preferential set indifferent to C_T. This innocuous-looking assumption (which requires that, in any continuous set, one cannot pass from preference to non-preference without passing through indifference) is actually of very great methodological significance in Georgesçu's epistemology. In fact, Axiom A must be stated as an assumption since the case of lexicographic preferences, for example, clearly show that it is not inherent in the structure of sensible preferences *per se*.

Georgeșcu continues by assuming no saturation region exists (Assumption B), limiting preference directions from any point are directly opposite (form a 180° angle, Assumption C), and that these limiting preference directions are, for any initial point, unique (Assumption D). Given these assumptions, individual preferences are given by the differential equation:

$$\sum_{i=1}^{n} \alpha_i(x)dx_i = 0$$

such that a move from x to $x + dx$, where the xs are bundles and αs functions, is preferred, non-preferred or indifferent as $\Sigma\alpha_i(x)dx_i >$, $<$, or $=$ to zero.

Georgeșcu proceeds to analyze preferences in terms of preference directions, and corrects an error in an earlier statement of "stability conditions in exchange" (essentially, diminishing marginal rates of substitution) by Allen (1932).[7] He also offers an interesting treatment of "saturation regions" in preference fields, and establishes that such regions must be convex. After some wandering, he establishes the main result, his first important contribution to economic theory: two points on the same integral surface may not be equivalent to the consumer, as the preference field may be "inconsistent". In particular, for nearby points to some bundle x, we may have $x \sim x'$ and $x' \sim x''$ but *not* $x \sim x''$, where "$\sim$" signifies indifference. In fact, what is missing is "transitivity" of the resulting indifference elements. This important insight finally brought to a close the "integrability puzzle" that had occasioned a considerable and somewhat fruitless debate among early mathematical economists.

Volterra, in a famous exchange with Pareto, claimed that the differential equation given could be integrated if and only if there were only two goods. Pareto argued that the differential equation given was equivalent to a (very large) set of observations on market behavior as described above. Yet, why should Pareto's idea fail if there were three (or more) commodities? Georgeșcu's results show that both great analysts missed the real difficulty: that integrability is not sufficient to establish an "ophelimity index." An extensive history of this debate is offered by Georgeșcu in "Vilfred Pareto and his theory of ophelimity" (1973).

Although "The pure theory of consumer's behavior" (1936) did much to establish Georgeșcu's neoclassical credentials, a careful reading reveals a wealth of references, "hints" really, to interesting ideas that were to later become central to his epistemology. First was the "ordinalist fallacy" concept of Assumption A. Second, Georgeșcu claims that dynamic and static models of consumer behavior differ primarily due to their mathematical representation. In particular, dynamic models do not admit treatment in terms of "Dirichlet" type functions, that is, single-valued functions whose values depend only on the point of evaluation rather than the path taken to obtain the point. This perhaps unsatisfactory view does suggest that, even as early as 1935, Georgesçu was beginning to wrestle with questions of hysteresis and "evolution," if only in a relatively limited way.

Georgesçu also argues in the "Pure theory", as he argued many times later, that consumer theory should be an analytical representation of how *facts* are organized and, further, how they "function." As he frequently noted, if a student asks a physicist what a symbol in some equation represents, in all likelihood the physicist could demonstrate with an experiment. Analytic models are useful only when they provide an accurate representation of the phenomena in question. What, then, of consumer theory? In Georgesçu's view, the greatest sin of neoclassical models of choice is not their artificial mathematical structure of infinitessimals and level sets, but the more fundamental error of assuming the entire problem of choice away by postulating a single source of satisfaction called "utility." All neoclassical choice models are seen as single-good models in which choice is little more than calculation. We will return to this idea below.

Georgesçu's preferred theory of "hierarchical choice" is treated later in this chapter, but the "Pure theory" shows a strong interest in the meaning of choice models very early in his career. Perhaps his background in statistics and relative "ignorance" of economics played a role in his scepticism. In any event, the "Pure theory" is full of both discussions of *testing* (some of) the postulates of choice theory and strong statements in support of introspective examination of the underlying assumptions of the models. The economic theorist is, after all, also a consumer who makes choices. Do the mathematical models used to represent choice correctly model this process?

This emphasis on "realism in modeling," in contrast to Professor Friedman's influential approach, was fully consistent with the discussion, common in the 1930s, on the "stability" of the consumer's optimum. Georgesçu, however, saw integrability itself as necessary (but not sufficient) for any choice process to be coherent. The uniqueness of indifference elements and transitivity of indifference were seen as critical so that the consumer, by some process of incremental adjustment, could "find" or "get to" the optimum point. Georgesçu frequently ridicules the idea, implicit in all modern neoclassical treatments, that this process of "getting to" an optimum point is somehow a separate issue to be ignored or relegated to some sort of independent search model.

Thus even in Georgesçu's earliest writings on utility theory we see a combination of technical innovation, rigor, concern for the *economic* meaning of assumptions, and emphasis on what the variables in the model actually mean. Given this, it is not surprising that Georgesçu made many other contributions to utility theory generally, often along novel lines. We turn now to an examination of a few of these contributions.

Threshold Phenomena in Choice

Georgescu's famous paper, "The pure theory of consumer's behavior", is actually (at least) two papers: the first deals with integrability and the nature of indifference curves, and the second is a largely independent, axiomatic model of stochastic choice, called a "threshold phenomenon" by Georgescu. It is a testament to his outlook that he saw the stochastic framework as, in fact, an elaboration of the general problem on non-integrability: in both the non-integrable and threshold problems, transitivity of the indifference element is lost. Yet the formal stochastic choice model introduced by Georgescu in the "Pure theory", and developed further in, for example, "Threshold in choice and the theory of demand" (1958) is of considerable independent interest. In this model Georgescu introduces a truly novel approach to consumer theory which contains, as a special case, the neoclassical model he did so much to perfect. Further, the analysis does much to highlight the important methodological insight, unrecognized in the 1930s, that choice models are ordinarily built on *binary* choices and that this assumption has strong implications whenever it cannot be assumed that consumers are "perfect choosers." Consumers could be called "perfect choosers" whenever they make decisions knowing perfectly both their preferences and the qualities and amounts of goods. This perfection is often called the "completeness" assumption in modern texts. When such superhuman perfection is lacking, the binary orientation leads to surprising dif-ficulties. For example, transitivity is lost, and the consumer, when faced with *many* (rather than two) alternatives, might select a bundle A, say, with the lowest probability although, given binary choices between A and any other feasible choice in the set, A may be the most likely choice in every binary comparison.[8] As Georgescu shows, the consequences of lack of perfection can be startling.

Georgescu's threshold model has the following structure. Given two bundles, A and B, in an n-dimensional Euclidean commodity space, the probability that the consumer will select A over B is W (A, B), and $W(A, B) + W(B, A) = 1$. Using this choice probability function, one can define several senses of "preferred to", such as "strong preference" ($W(A, B) = 1$), preference ($W(A, B) > \frac{1}{2}$), complete indifference ($W(A, B) = \frac{1}{2}$), indifference ($W(A, B) \neq 0, 1$), and so on. Next, several axioms are introduced, including (a) if $A_i > B_i$ for all goods i, then $W(A, B) = 1$; (b) $W(X, A)$ is continuous in X except at the point $X = A$; (c) if $A_i \leq B_i$ for all i, then $W(A, C) \leq W(B, C)$; (d) if $W(A, B) = W(B, C) = p \geq \frac{1}{2}$, then $W(A, C) \geq p$; (e) if $C = \lambda A + (1-\lambda) B$, and $0 < \lambda < 1$, then $W (A, B) \leq W(C, B)$. These axioms are generally quite intuitive: more is preferred to less; preferences are "pseudo- transitive"; preferences are convex in a specific sense.

These axioms, and a few other technical assumptions, allow the derivation of a large number of theorems on the structure of the preferential sets. For example, it is straightforward to show that upper-level sets defined by the relation $W(X, Y) \geq p$ for fixed Y are closed, convex, $n - 1$ dimensional manifolds. Further, the indifference varieties given by relations of the form $W(X, Y) = p$ for fixed p and Y exhibit a one-to-one correspondence with the real line. It is also possible to show that, for any three bundles A, B, C, then $1 \leq W(A, B) + W(B, C) + W(C, A) \leq 2$.

Georgesçu proves that his axioms are internally consistent and, under suitable restrictions, are consistent with the usual neoclassical axioms of utility theory. But he does far more. In particular, pick a bundle A and consider the family of indifference curves generated by bundles X satisfying a condition of the form $W(X, A) = p$ for various values of p. This allows us to define two regions, emanating from A, that are regions of "thresholds" which contain bundles X such that $W(X, A) \neq \{0, 1\}$.

Formally, these regions are the solutions X to the inequality $[H(X;1) - H(A;1)] [H(X;0)] < 0$, where $H(X;p)$ is the integral of $\alpha_1(x; p)dx_1 + \alpha_2(x, p)dx_2 = 0$, that is, the p indifference curve. These threshold regions are very small near A but become larger for bundles X far from A. This implication of the axiomatic framework is not obvious on initial examination, but is a very interesting result.

Having established the threshold framework, Georgesçu examines its implications. First, he presents a set of axioms on the probability functions $W(A, B)$ generalized to many choices, the interpretation of the expression $W(X_1, X_2, ..., X_n)$ being the probability that the order of preference among the Xs is $X_1, X_2, ..., X_n$. Next, he shows that two sensible-sounding axioms are then incompatible, a technical result that establishes the following possibility. Suppose the bundle A is a "tangency point" given the consumer's budget line and the indifference curve $W(A, X) = \frac{1}{2}$. Then it is *not* true that A must be the most frequently chosen bundle when all possibilities are considered together. This strong conclusion highlights the implications of binary choices with threshold phenomena. Binary choice cannot be the basis of demand theory unless people are "perfect choosers." Tangency points need not be the most frequently selected bundles, thus one cannot argue that neoclassical choice theory represents "average" behavior. As Georgesçu points out, threshold phenomena lead to demand "distributions" where, given some price system, demand quantities are stochastic, and the bundle representing the tangency of budget plane and the strict ($p = \frac{1}{2}$) indifference curve is not necessarily the most probable choice.

Hysteresis in Preference

Georgesçu-Roegen believed that the economic process was an historical, evolutionary one involving irreversible, non-arithmomorphic changes at its very heart. Further, these changes take place at both the micro and macro levels of the economy.[9] In an interesting illustration of the consequences of accepting this commonsensical view, he proposed a theory of consumer choice in which "history matters," described in his paper, "The theory of choice and the constancy of economic laws" (1950).

Georgeşçu begins with a critique of the axioms of standard choice theory. Ordinarily, preferences are considered to obey a small set of axioms familiar to students of microeconomic theory: preferences are complete, more is preferred to less, and preferences are transitive. One then begins an investigation into the shape of indifference curves. Yet, as Georgeşçu argues, this appears to be a classic case of "jumping the gun." If one interprets utility theory as describing "average" behavior, some demonstration must be made that the neoclassical optimum, a budget plane/indifference curve tangency point, identifies a bundle with the greatest likelihood of being chosen by the consumer (or is "average" in some other suitable sense). Yet, as shown in the last section, there is no requirement that such a result be obtained: binary choice analogies may be grossly misleading when choice is imperfect.

Second, Georgeşçu questions the process by which a consumer, starting at some endowment, actually arrives at an optimal choice. Is it a process of trial and error? Or is the consumer, in Georgeşçu's telling phrase, a "homing pigeon?" It is important to recognize that this issue, largely ignored in modern neoclassical presentations of choice theory, was considered central to discussions in the 1930s. Instead, the notions of equilibrium "stability" (diminishing marginal rates of substitution), analyzed extensively (and incorrectly) by R.G.D. Allen, were seen as vital to any explanation of how the equilibrium point was obtained. Georgeşçu himself argued that integrability was important because it had direct bearing on the feasibility of "local" adjustments leading a consumer from an endowment to an optimal point.[10] Finally, it is also apparent that this issue is logically distinct from any problems of measurement of goods or their uncertain qualities.

Third, Georgeşçu asks what role the consumer's *experiences* have in shaping the preference map. Although it would seem a commonplace that experience affects preferences, no mechanism for this interaction is apparent in conventional models. Preferences are taken as "given," as if such a phrase somehow solves the problem. Arguing that the process of preference formation is noneconomic, that is, independent of consumption choices, seems incredible. The importance of completeness axioms is apparent. Becker's work on habit formation is a response to these concerns.[11]

Georgeşçu clearly believed that experience affected preferences, which in turn led to selections of new consumption bundles, which led to further changes in preferences, and so on. Besides the rather convincing evidence for this view available from simple introspection, Georgeşçu was impressed by evidence, associated with debates on consumption functions and the relative income hypothesis of Professor Dusenberry, that an income change, even if transient, could lead to a permanent alteration in consumption patterns, implying changes in the underlying preference field. Although one may today question the econometric material on which this finding was based, the notion of preference "evolution" was completely consistent with Georgesçu's economic methodology in general.

Georgesçu restricts his attention to the existence of "indifference varieties" (indifference curves), and does not consider their shapes. He incorporates hysteresis into choice theory by his customary axiomatic approach. Let c_i be bundles, where bundle c_i is experienced by the consumer with time duration t_i. Then let $P(c_1, c_2, t_1, t_2)$ = probability that the consumer selects bundle c_1 over bundle c_2. A variety of postulates are proposed which are similar to those offered in his analysis of threshold phenomena. Now, however, the time duration element necessitates some new assumptions. In particular, Georgesçu proposes that $\lim P(c_1, c_2, t_1, t_2)$, as t_1 and t_2 both go to infinity, is either 0, ½, or 1. Further, if the consumer is indifferent between c_1 and c_2, then $P(c_1, c_2, t_1, t_2) = $ ½ for all $t_1, t_2 > 0$.

Georgesçu's final postulate, the "Hereditary Postulate", holds that *history matters*. The consumption history of individuals affects their indifference maps. Let S be the set of all bundles previously tried. Then, ignoring time, an indifference curve has the form $\rho(c;s) = $ constant, for some function $\rho(\cdot)$ which is a point function of c but a set function (though not set valued!) of S. Any choice c can fall into (at least) three categories. Either c has been tried and has no effect on preference, or else c has not been tried but will have an effect, or c has not been tried but would have no effect. Thus, any bundle consumed is either an ineffective repetition, an ineffective "novelty," or a "new relevant experience."

Georgesçu's examples and discussions suggest he had the following ideas in mind. First, since the time element is here being ignored, a "resampling" of a previous bundle will have no effect: recall is perfect. Second, a "new" bundle which is "similar enough" to previously tried bundles can have no effect. Third, certain geometric relationships between bundles in the goods space determine the effects of consumption.

Let $\bar{S}$ be the convex hull of S and a few additional, prespecified bundles, that is, $\bar{S}$ is the set of all bundles obtainable by convex combinations of all previous bundles plus a few others. To represent the consumer's inherent knowledge, let $\bar{S}$ include the additional, unsampled bundles composed of, say, a set of basis vectors for the commodity space. Georgesçu illustrates his model for two goods X and Y by supposing that preferences are given by:

$$\ln\rho(c;s) = A(\bar{S})\ln X + B(\bar{S})\ln Y$$

where $A(\overline{S}) = (1 + a) / (2 + 3a)$, and $B(\overline{S}) = (2 + 5a) / (1 + 2a)$, where a is the *ratio* of the *area* of $\overline{S}$ to that of a triangle in the plane formed by the basis vectors and the origin. As new relevant experiences occur, they are incorporated into an enlarging convex hull $\overline{S}$, causing a to increase, thereby altering the preference map. We have $\lim A = \frac{1}{3}$ as $a \to \infty$, and $\lim B = \frac{5}{2}$ as $a \to \infty$, which presumably represents the "true" preferences in the sense that these preferences would prevail given "infinite experience" by the consumer.

Several points need to be made about this example. First, it is only a very contrived example, although it serves well to illustrate, at the simplest level, what Georgesçu intends. Note first that this model leads to irreversibility of demands: any price change that induces selection of an effectively novel bundle, if reversed, would not return the consumer to his or her previous equilibrium. Second, there is no place in this analysis for bundle selection based on some forecast of the resulting effects on preferences. Of course, the preference adjustment process given in the example is mechanical, so one could presumably foresee all the future consequences of any selections of bundles. Yet there is no basis in the model for the agent to prefer one set of preferences over another, and such an idea, which may be senseless if preferences are correctly defined, would presumably require some "meta-preferences" (whatever that may mean).

In any event, Georgesçu is quick to point out the consequences of his approach for such essential topics as demand estimation and econometric modeling. For example, if price changes are small, then only small consumption changes occur (that is, when preferences are strictly convex) which, while negating preference alteration, gives little variation in the data with which to construct reliable statistical estimates. Alternately, if price changes are large (as the econometrician might hope), the probability of a large consumption swing altering preferences is high. Thus when the data are good, the demand relationship is not stable, and when the data are bad, the estimation is unreliable. This rather resembles a "Heisenberg Uncertainty Principle" for econometrics.

Among the most interesting parts of Georgesçu's discussion of hysteresis is his digression on the notion of "equilibrium" in a model in which consumption decisions alter preferences. He notes that, under such circumstances, "equilibrium" could mean either: (a) a position foreseen prior to experimentation; (b) a position that satisfies some equilibrium conditions if hit upon at once; or (c) a position that satisfies some equilibrium conditions if obtained by trial and error. This discussion, though taking place in 1950, is exceedingly modern in spirit and foreshadows important current ideas in rational expectations and refinements of solution concepts in game theory. In particular, Georgesçu's equilibrium type (b) resembles the Nash solution concept, with the mutual best reply property serving as the "equilibrium condition" if hit upon at once, while (c), stressing the "extensive form" (sequence of moves), suggests a learning process. This analogy is not exact, of course, since at a minimum Georgesçu views the set of points which could potentially be equilibria as dependent on the adjustment path followed, and, in any event, a decision problem is not much of a game.

Georgesçu's Model of Hierarchies of Wants

An introspective examination of buying and consumption behavior can put neoclassical utility theory in a rather peculiar light. Consider, for example, any principles-level textbook treatment of the notion of "diminishing marginal utility." Explanations of this principle rely on the notion that wants, which are satisfied by consumption, are so satisfied in a *sequential* fashion. Yet the existence of this sequence implies that wants have some sort of hierarchical structure, and that consumption even of multiple units of a single good produces a stream of pleasurable impulses which, in some qualitative sense, are diminishing.

Although it is widely believed by neoclassical economists that the modern "theory of choice," relying as it does on a diminishing marginal rate of substitution (convexity) principle instead of any notion of marginal utility, has successfully dispensed with the very notion of "want," it may be fairer to say that all that has occurred is a replacement of the reality of many wants with the fiction of a single want: utility. Our desires for comfort, variety, and the like are all subsumed under the rubric of "utility maximization." This result is achieved at a cost in both realism and abstract mathematical structure.

Georgesçu, whose personality and inclinations compelled him to closely examine what most economists take for granted, provided an extensive explanation of the role of "wants" in choice theory in "Choice, expectation, and measurability" (1954). In particular, he felt that ignoring the hierarchical structure of wants led economists to commit a fundamental error with profound implications for demand analysis. This error arose from an "ordinalist fallacy": a property such as continuity which applies to a variable used to define states (such as "more preferred") is erroneously assumed to apply to the states themselves. To see this, consider the following circumstance. A person has two wants: to be free of thirst and to have a well-maintained lawn. Water, a commodity, is effective in satisfying both wants. Georgesçu asserts that this person may well drink water to the satiety of thirst (so its marginal utility in this application is zero), while watering the lawn to some point below satiety in that application (so its marginal utility is also positive). A neoclassical economist, confronted with this parable, must claim that the facts as stated are in error: the marginal utility of water per dollar must be the same in all its applications. This result arises from utility maximization. Alternately, a neoclassical may assert the existence of other costs associated with lawn watering.[12] Yet this response is merely a claim that the behavior described by Georgesçu is impossible. It is certainly unclear why a mathematical theory describing human behavior could become a law governing it. In any event, Georgesçu saw the textbook treatment as suspicious.

In order to illustrate the implications of the hierarchy of wants on consumer behavior, Georgesçu offers a set of four "principles":

1. *Principle of Subordination of Wants*: the satisfaction of a lower want allows a higher want to manifest itself.
2. *Principle of Satiable Wants*: all wants can be satiated.
3. *Principle of the Growth of Wants*: the satisfaction of one want always leads to another, so wants are unlimited.
4. *Principle of Irreducibility of Wants*: wants are irreducible, so the satisfaction of one cannot eliminate another.

Georgesçu envisions a hierarchy of wants beginning with thirst, hunger, shelter and similar categories, and progressing "upwards" to include higher level wants which are culturally motivated. Such a hierarchy is assumed to apply to everyone. There is no simple, one-to-one correspondence between wants and goods. Choice is the act of satisfying wants, beginning with the most fundamental. Thus a choice between bundles reduces to a comparison of satisfactions of the least important wants.

Although this discussion appears somewhat vague, Georgesçu has something very specific in mind, and he illustrates his ideas through an extended example that it is worthwhile to reproduce. A consumer buys two goods, butter X_2 and margarine X_1. His wants, in relevant order, are for (a) calories $k = X_1 + X_2$; (b) taste $t = X_2$; and (c) companionship of friends. Up to some point $k = \bar{k}$, his selection is solely based on calories. If two bundles yield equal calories less than $\bar{k}$, he selects lexicographically on t. For $k > \bar{k}$, he selects on t alone. If both bundles have $k > \bar{k}$ and yield the same t, he goes to his third want, and so on. Thus this consumer prefers $c' = (x_1', x_2')$ to $c'' = (x_1'', x_2'')$ if either (a) $k'' < k' < \bar{k}$, or (b) $k' = k'' \leq \bar{k}$, $t'' < t'$, or (c) $k'' < \bar{k} < k'$, or (d) $\bar{k} < k'$, $\bar{k} < k''$, $t'' < t'$, or (e) $\bar{k} < k'$, $\bar{k} < k''$, $t'' = t'$, $e'' < e'$, where e measures hours of companionship.

What do such preferences imply? First, such preferences satisfy the neoclassical postulates as usually stated, but not Assumption A of Georgesçu's 1936 paper, "The pure theory of consumer's behavior." As Georgesçu showed, there is no ordinal measure of utility (no utility function) for this consumer even though all bundles can be unambiguously ranked. It is impossible here to place the bundles in correspondence with the real line. This significant complication is omitted by neoclassical theory. Yet if one accepts Georgesçu's hierarchical formulation, the significance of this issue is apparent: consumers do not have utility functions because such a representation is inconsistent with their preferences.

Second, it is apparent that the indifference element does not exist. We "pass" from preference to non-preference without crossing through a state of indifference. If this offends our sensibilities, it is only because we are assigning the properties of a variable describing states to the states themselves. There is no reason to imagine that the topological structure of preferences is isomorphic with that of the Euclidean space. In fact, Georgesçu argues they are quite unalike.

Finally, our imaginary consumer may present policy makers with some interesting challenges because he or she does not satisfy the "Principle of Substitution." Thus the consumer cannot be compensated with one good to make up for the lack of another, in the same way that a person dying of thirst cannot be compensated for a loss of water with loaves of bread. Unless the bundles are identical, the consumer will not be indifferent between them. This result, of course, is built into the example, and it is possible to modify it with "redundant" products to allow for limited substitution. The example remains a worthwhile antidote to any unwarranted optimism about the perfection of neoclassical models of preference.

Georgesçu and Expected Utility Theory

Although most of Georgesçu's writings on consumer choice represent critiques of the neoclassical choice under certainty model, he had great interest in the issues of uncertainty, expectation, and probability. His background in statistics no doubt played a role in this, and he clearly saw choice under certainty as merely a special, limiting case of the general choice problem. Unlike many critics of expected utility theory, such as Professor Allais, Georgesçu did not object to von Neumann and Morgenstern's "cardinalist" model because of the independence axiom and its supposed violations.[13] Rather, Georgesçu felt that the flaws in expected utility theory arose primarily from the purely ordinalist axioms it inherited from the certainty theory.

Georgesçu held different views on the distinction between risk and uncertainty at various points in his career. In the 1950s, his views were much influenced by the well-known definitions proposed by Frank Knight: risk describes the circumstance in which the outcome of a process is unknown because the outcome is generated by a known process but under an unknown initial condition, while uncertainty refers to the case where the process itself is unknown. Thus "probability" is really a *physical* characteristic of a mechanism and applies only to cases of risk. In later writings, such as *The Entropy Law* (1971), Georgesçu broadened his views on uncertainty considerably.

The conventional axioms of expected utility theory are: (a) completeness; (b) non-satiation; (c) transitivity; (d) some convexity assumption; and (e) the independence axiom (if two lotteries are blended with a third one, then any preference between the resulting compound lotteries is unaffected by the third lottery selected).

Georgesçu forcefully argued that the ordinary axioms of utility theory are incomplete without the addition of some postulate corresponding to his Postulate A ("The pure theory of consumer's behavior," 1936) to assure existence of ordinal measurability of preferences.

Georgesçu, in contrast to many critics of expected utility theory, had no quibble with the independence axiom *per se*. This axiom (plus the others) yields (weakly) cardinal utility, and it reduces the comparison of all risky prospects to a comparison of expected utility, a scalar. Yet can the first moment (expected value) of utility be a valid description of how risky choices are made? Georgesçu says "no," but *not* because of the independence axiom. Rather, the need to assume ordinal measurability is the critical error. For example, he suggests that a risky prospect can *never* be equivalent to a riskless one: there are not enough points in the continuum to associate a different number with each different uncertainty.[14]

To explain the source of Georgesçu's complaint, consider the following example. An urn is filled with N balls, *m* of them white and *n* of them black. $P = m/N$ is the true probability of drawing a white ball, while some P' is an estimate of this probability generated by a sampling of size $\tilde{N}$. Thus P' is a measure of P, and $\tilde{N}$ an indication of our credence in it. Uncertainty, then, is taken to refer to both the estimate of P and our reliance on it.

Suppose one is offered two urns. Using a sample of size $\tilde{N}_1$, the first urn produces p_1, while on a sample of $\tilde{N}_2$, the second produces an estimate $p_2 \leq p_1$. Suppose further that $\tilde{N}_2 \leq \tilde{N}_1$. One is asked to select an urn so one will receive a prize if a white ball is picked. Which urn is selected? The critical issue here is the capacity of the model to represent "tradeoffs" between probability and credibility. Obviously, for the same level of credence ($\tilde{N}_2 = \tilde{N}_1$), anyone would select the urn with greater p'. Alternately, for the same probability estimates ($p_1 = p_2$), anyone would select urn 1. Things are far less clear when $p_1 < p_2$ and $\tilde{N}_1 > \tilde{N}_2$. Reasonable people, confronted with this circumstance, may well have preferences which are *not* ordinally measurable, and indeed we use such preferences in econometrics frequently: best linear unbiased estimators (BLUES) are minimum variance among a class of unbiased estimators linear in the data.[15] Tradeoffs between variance and bias are not allowed.

The importance Georgesçu attached to ordinalist fallacies in the analysis of uncertainty and expectation led him to return to this theme many times. Perhaps his most developed statement is offered in "The nature of expectation and uncertainty" (1958). In this essay he provides probably his strongest arguments for the inadequacy of real numbers to express expectations. Further, he articulates quite clearly his opposition to both subjectivist and pure frequentist doctrines of probability. Several points are worth emphasizing. First, the claim that ordinal measurability problems afflict expectations is entirely consistent with Georgesçu's analysis of choice under certainty, and is complementary to it. Second, one can view all his work on utility theory, much of which was completed by the 1950s, as a prelude to his most important methodological point, the inadequacy of mechanical models in representing economic phenomena. Viewed in retrospect, Georgesçu's proposals for improving utility theory represent responses to the complaints that, in neoclassical models: (a) consumers somehow measure goods and their qualities perfectly (threshold phenomena); (b) consumers' choices do not affect their preferences (hysteresis); (c) neoclassical consumer choice is a process that combines a set of feasible elements with a maximization rule (utility function), and thus is merely a mechanical process (hierarchy of wants and ordinalist fallacies of expected utility theory). Thus, Georgesçu positions himself as that most welcomed element among the set of least popular characters: a critic with an alternative in hand. That Georgesçu's alternatives are mathematically difficult is obvious, but they are alternatives nevertheless.

4.3   GEORGESÇU AND PRODUCTION THEORY

Georgesçu's early interest in utility theory was soon matched by an interest in the complementary problem of modeling production. Although most of his publications in production theory date from the post-Second World War period, there is considerable evidence that his interests predate the war and were in fact established during his time at Harvard in the 1930s.

Although it could certainly be argued that production is, in fact, at the heart of much of Georgesçu's later work on resource economics and the Entropy Law, he made several important contributions to production theory within the "normal science" interpretation of that term, and it is these works that command our attention in this section.

First, in the late 1940s, as perhaps his first post-Second World War economic theorizing, Georgesçu began a careful study of the Leontief linear model of production. With the publication of Leontief's famous book, *The Structure of the American Economy, 1919–1929* (1941), this model became the focus of much research by Samuelson, Dorfman, Solow, and many other leading economic theorists and mathematical economists of those times. This model, which most modern students of economics encounter in simple examples in primal (goods) form, was actually introduced by Leontief in money terms, and held the promise of being an estimable, practical model of national economic activity, consistent with Walrasian general equilibrium conditions. The policy value of such a program is, even today, quite obvious. Thus, as many leading economists did, Georgeşçu began seriously to evaluate Leontief's path-breaking contribution and, to Georgeşçu's credit, he promulgated several important results. His presentation of these results, most of which appear in his papers included in Koopman's *Activity Analysis of Production and Distribution* (1951), constitute one of Georgeşçu's most important contributions to mainstream economics.

Somewhat later in his career, Georgeşçu began to think more deeply about the epistemological status of neoclassical models of production and to formulate a highly original analytic representation of production, often called the "flow-fund" model, which he described in his 1969 Richard Ely lecture. This model was developed further for inclusion in *The Entropy Law and the Economic Process* (1971), and it is this latter presentation which forms the basis of our discussion here.

Non-substitution Theorem

Georgeşçu, perhaps contemporaneously with Professor Samuelson and some others, discovered an important theorem in the basic Leontief model which, while usually referred to as the "Substitution Theorem," should more precisely be called the "Non-substitution Theorem," as it provides a somewhat surprising justification for the restrictive specification of technologies in the Leontief model. In order to explain this theorem, it is necessary first to provide some background on the Leontief models familiar to Georgeşçu. This groundwork offers the additional advantage of bringing to the fore several themes that would preoccupy him in some of his later work on development and production models.

The "Leontief model" is actually an umbrella term for a large number of closely related models which differ in some fundamental ways. Closely related to these models is the von Neumann model (1945), about which Georgeşçu also wrote extensively.[16] For our purposes here, we can begin with a simple version that highlights some of the primary conceptual issues.

The Leontief model is an input–output model which illustrates the interdependencies of production among different industries, though these industries are often denoted as "activities." All goods are produced by a combination of inputs including a primary input, labor, and other outputs used as inputs. For example, steel is produced by labor, iron, coal, electricity, and so on. In turn, iron, for example, is produced by labor, coal, electricity, *and* steel. Ordinarily, all outputs use many or all other outputs as inputs. Labor, the primary input, has a special status, and may not be produced in the sense that the other inputs/outputs are. This view of production differs substantially from the vertically structured, "ordered" framework associated with the Austrians and some classical economists.[17] No industry is viewed as "prior" to any other.

Production of good i, X_i, is decomposed into inputs of i into all industries, including industry i, where X_{ij} denotes input from activity i into activity j. If C_i is the final consumption of good i, then

$$X_i = C_i + \sum_j^n x_{ij}$$

where n denotes the number of activities. Labor services are inelastically supplied in quantity X_0. Conventionally, labor is not consumed ($C_0 = 0$) so that

$$\sum_i X_{0i} = X_0.$$

Isoquants are assumed to be convex, and constant returns to scale are assumed to prevail. Finally, and most characteristically, Leontief assumes that production functions can be written as $X_i = \min (X_{11}/a_{11}, X_{21}/a_{21}, ..., X_{01}/a_{01})$ for produced goods X_i, where the a_{ij} are fixed coefficients of production. Then the matrix $[a_{ij}]$ summarizes the technical requirements for production, with a_{ij} indicating the per-unit requirement of good i in producing good j. Clearly we require $a_{ii} < 1$ for "production" to occur at all. Thus, isoquants are right-angled, evenly spaced and, in the absence of additional assumptions on capacity, extend outward forever.

It is important to note some of the limitations of this approach. First, each sector of the economy produces only a single good: joint production is ruled out. Second, substitution among inputs is severely constrained although, given the existence of a single primary factor (labor), substitution is rather a misleading term anyway, as will be explained below.

A first question naturally concerns the set of feasible final demands C, or, in the common nomenclature of the time, feasible "bills of goods." Much research went into the solution of this problem, and Georgesçu extended some of this analysis in his paper, "Some properties of a generalized Leontief model" (1951). The celebrated Hawkins–Simon conditions (Hawkins and Simon, 1949) provide an answer stated mathematically. The ordinary language equivalent is: to produce one unit of X_1, say, it is necessary that the direct requirement a_{11}, plus the "indirect" requirement for X_1 embodied in all other inputs, be less than one unit. This condition is more complex and deeper than it initially appears, because on reflection it is clear that it involves the limit of an infinite series of requirements: to produce X_1 one needs some X_1 directly, then some X_1 embodied in other inputs, then further X_1 embodied in inputs used to produce other inputs, and so on *ad infinitum*. This presents no purely technical problem since simultaneous solution addresses the infinite regression. However, as a practical matter, the existence of these dependencies creates a very interesting problem in interpreting how an economy with production described by the Leontief system is supposed to grow through time. As Georgesçu noted, an increase in X_0 (labor) would seem to be the only source of growth in such an economy in the absence of technical change (a shrinking of the a_{ij}s). In fact, Georgesçu attributed much of the West's economic growth to a substantial lengthening of the working day.[18]

We can now state the famous "Non-substitution Theorem" established by Georgesçu. It is clear from our previous description of the Leontief system that one restriction imposed on production is that every sector (activity) has only one technical recipe for the production of output. Although this would seem to be a very strong requirement, the "Substitution Theorem" (or, more properly, "Non-substitution Theorem") suggests that this may be unimportant. Thus the Substitution Theorem establishes a level of generality for Leontief-type analysis that increases the importance of these models.

Although several forms of this theorem are presented in the essays in Koopman's famous book, the basic insight is the same in each case. Given the structure of the simple Leontief system, which has only one primary input (labor), the observed input ratios in all processes will be invariant with respect to the structure of final demands (the Cs) even if each industry has several production processes available to it. This result, which seems so inconsistent with ordinary neoclassical intuition on substitution, arises because, if labor is the only primary input, then the cost of any good can be reduced solely to the cost of its direct and indirect ("embodied") labor content. Cost minimization then implies that the social costs of goods, which are just the value of their labor content, be minimized. Unless two processes utilize identical labor contents for the same good, one must always be cheaper than the other regardless of the levels of final outputs that arise from either central planning or some market process. Changes in wages lead to proportional changes in the costs of all goods. The simplest proof of this proposition involves construction of the convex hull of all available "simple" processes: the resulting production possibilities surface is still a straight line and is constructed of weighted averages of single processes or of multiple processes having the same technical coefficients for all activities.

This result clearly arises from the structure of the Leontief system and, in particular, the restrictive assumptions imposed in the simpler of such systems. Joint production, in its real sense, may very well alter this conclusion. Economies of scale will also generally alter the conclusion since, in that case, changes in the scale of operations will change the embodied labor content of goods and their costs in the one primary factor model.

Georgesçu and the Flow-fund Model of Production

Among Georgesçu's most original contributions to economics is his model of production, usually termed the "flow-fund" model. As with so many of his contributions, the flow-fund model grew out of the dissatisfaction he felt toward the rather cavalier treatments of the complexities of economic life by neoclassical economics. As with the typical formulations of utility theory, Georgesçu's investigation into the epistemological underpinnings of the "production function" led him to conclude that the ordinary treatment of production offered by mainstream texts was seriously flawed, ignored critical aspects of the production process, and produced misleading, or even perverse, conclusions.

It appears that Georgesçu developed what we now call the flow-fund model slowly, over many years. However, by the mid-1960s many of the characteristic elements of his approach were formulated. Although some discussion of the insights underlying the flow-fund approach appears in the introductory essay of *Analytical Economics* (1996), Georgesçu's Richard Ely lecture of 1969 provided the first comprehensive (though not comprehensible) explanation of the approach to a wide audience. The description provided in the Ely lecture is expanded into a long section of *The Entropy Law and the Economic Process* (1971), and it is from this source that what follows is mostly drawn.

To begin, it is important to review the rather poorly articulated neoclassical treatment of production which motivated Georgesçu to look more deeply. A common statement of the production relationship might be: "Let Q represent the product, and K, L, M ... the factors of production, then the product obtainable from the factors is summarized as $Q = f(K, L, M ...)$ for some non-decreasing point function $f(\bullet)$." The term "point function" refers to a function whose range is the real line. Ordinarily, it is left to the reader to decide if Q is a rate of production, or a stock of output on hand at the end of some specified time. Likewise, the "inputs" K, L, M ... are usually treated as flows of some sort, although this is often not made explicit.

As early as 1949, Georgesçu had called to account the logical consistency of the neoclassical formulations, particularly those of Leontief and von Neumann. Because these criticisms are quite relevant to later work on the flow-fund model, they are worth reviewing at this point. Often, the neoclassical production function is termed a "point" production function for which the time element is omitted entirely, as when $Q = F(K, L, M, ...)$ is specified such that Q, K, L, M, and so on are all quantities (stocks). Alternately and/or strangely often simultaneously, economists write $q = f(k, l, m ...)$ where all magnitudes are now interpreted as flows per unit of time. Georgesçu, in one of his most celebrated but simple proofs, showed that, if both representations $Q = F(K, L, M ...)$ and $q = f(k, l, m, ...)$ are correct, then we must have $F = f$ and homogeneity of degree +1. This strange result arises quite simply from the identities $K = t\,k$, $L = t\,l$, and so on, for any time interval t. Georgesçu was fond of claiming that he had in fact "proved" that all production processes exhibit constant returns to scale. Of course, what he actually showed was that (a) the neoclassical treatment was sloppy, and (b) you can't have it both ways. In passing, Georgesçu claimed that this result astounded a prominent colleague who, in response, refused to comment on Georgesçu's paper due to its "fundamental mathematical error."[19]

Georgesçu's criticisms serve to identify the two primary analytical representations of production "processes" used in mainstream economics. The first, identified with Leontief, holds that a process is completely described by its flow coordinates. Production is identified with flows across an imaginary boundary. Alternately, von Neumann identifies a production process with two "snapshots" (in Georgesçu's terms), each a census of all the production factors (inputs and outputs) at two times t_0 and t_1, $t_0 < t_1$. In this view, we need only take count of the *stocks* of commodities present at t_0 and t_1.

Georgesçu finds both conceptualizations of production to be wanting. The flow model, for example, ignores the existence of productive agents, such as Ricardian land, which reside inside the process boundary and are not associated with across-boundary flows. These agents are of many kinds and we will return to this theme again below. The "stock" model, in contrast, is inherently incapable of representing change in any real sense: if the stocks of productive factors are the same at t_1 and t_2, are we observing a steady state or a "frozen conglomerate?"

All these complaints probably seemed strange to the neoclassical economists: Could one not move between flow and stock models via the "identity" $\Delta S = S(t_2) - S(t_1)$, where ΔS is the "flow," $S(t_i)$ the stock at time i, and $t_2 \geq t_1$? Encapsulated in this simple suggestion is the idea, crucial to neoclassical models of economic dynamics and anathema to Georgesçu-Roegen, that "change is locomotion." Stock and flow formulations are neither equivalent nor economically correct. As Georgesçu noted, it is not true that "the census taker must come out with exactly the same list of elements as the customs officials..."[20] A flow does not necessarily imply the diminution of a stock of the same substance: time flows but neither arises from or drains into any stock. Output itself, which has a flow representation, does not arise from depletion of an actual stock, but is instead created in production.

But stocks and flows are related in the sense that flows can be conceptualized as *stocks spread over time* although, in some cases, the stock in question is "analytical" only. Put this way, it is clear that inputs used in economic production may be conceptually mutilated by the stock/flow equivalence relied upon in many neoclassical production models. To highlight this distinction, Georgesçu introduces the idea of a "fund factor," or "fund of services." At the beginning of the production process, a production unit, such as a factory, will have on hand various productive agents, including, perhaps, fuel and some piece of capital equipment, such as a furnace. There is, however, a very important distinction to be made between these factors. The fuel exists at the initial time t_1 as a stock, and this stock will be decumulated over time as a flow of fuel to fire the furnace. Yet the furnace does *not* represent a "stock of services" because, unlike the fuel, it takes a fixed time to decumulate it. In other words, although, in theory, one could use up all the fuel in any finite interval of time, regardless how short, "using up" the services of the furnace must proceed at an absolutely constant (maximum) rate that may well require years for exhaustion. Thus one cannot correctly speak of a *flow* of services from a fund factor. Many capital equipment services and labor services are, in fact, funds. To highlight this distinction, consider the dimensions of the units used to measure flows and funds. The amount of flow (of, for example, fuel) is in units such as pounds, while the rate of flow will be, for example, pounds/week. In contrast, the amount of services of, for example, a furnace will be in furnace hours (furnace × hours), while the *rate* of services is just *furnaces*. This distinction is far from being a mere academic quibble, as will be seen below.

The distinction between flows (stocks spread over time) and services of a fund has crucial implications, according to Georgesçu, for the market pricing of these elements. The non-existence of a "spot" market for fund services ordinarily implies the unintended idleness of these funds over some part of the production period. Since the services of human agents are likewise funds, humans may experience undesired idleness, and such phenomena formed an important reason for Georgesçu to deny the efficiency of marginal pricing in "underdeveloped," agrarian economies. Indeed, he even attributes recessions to the economic importance of fund factor decumulation constraints.[21]

By fund factors, Georgesçu ordinarily means capital goods, labor "power," and land in its Ricardian sense. Flows are taken to represent everything else, especially natural resources (including solar radiation), material flows, maintenance flows, and waste outflow. The final flows of interest constitute the output goods or products that are the aim of production.

We turn now to the analytic representation of the production process. Although a process is, in fact, a "happening," it is necessary first to define a "stationary state," fictitious as it may be. Growth, for example, is the passage from one stationary state to another. A "stationary state," for Georgesçu, is a state of a system that can be repeated over and over, for example, Marx's "simple reproduction" would qualify. Yet such a status obviously requires that fund factors "inside" the analytical boundary of the process be somehow maintained in their original conditions. Georgesçu illustrates the quite hypothetical character of this requirement by the example of farming as a process. The farmer uses a shovel. To maintain the shovel he requires a file. To maintain the file he requires additional equipment, and so on *ad infinitiun*. This example also shows that farming is indeed a process: one test for true "processhood" is the emergence of infinite regressions.

Thus, we arrive at the following conceptualization. The production process is a "cut" in the seamlessness of reality. The process is a happening delineated by an analytic boundary across which we measure the passage of resources over time. We invoke the fiction of "maintenance flows" across this boundary which maintain the fund "agents," located "within," in their original condition. Everything in the universe is either the process or its environment: the boundary has no penumbra.

Broadly conceived, the elements of any process are of three types. First, we have those elements which may enter or leave ("cross") the analytic boundary. These are cardinal magnitudes, and some may go only in one direction, whereas others may enter and leave unchanged. Second, we have elements, such as the economist's "Ricardian" land, which yield services but are not altered by doing so. Finally, some elements clearly undergo a qualitative alteration which cannot be accurately represented by real numbers: a worker goes from rested to tired. Because analytic representations of qualitative changes are inherently impossible, and because analysis requires such a representation, Georgesçu typically adopts the convention of using two coordinates to represent each such factor.[22]

Lacking from our description thus far is the notion of time. As explained previously, duration is, for Georgesçu, a critical requirement for any process. In fact, a specification of a duration, a starting and ending time, is inherent in the definition of process. Thus, let us call the initial instant at which we begin to study the process 0, and the final instant at which we stop T. Then we have a representation of a process as:

$$\left[\underset{0}{\overset{T}{E}}(t);\ \underset{\alpha}{\overset{T}{S}}(t) \right]$$

where $E_i(t)$ represents the cumulative (cardinal) amount of flow factor i at time t, while $S_\alpha(t)$ shows similarly the amount of fund services rendered up to time t.

Whether a factor is a fund or a flow element is at least partially time dependent: as T becomes large, entire pieces of equipment, for example, are eventually replaced, rendering them flow factors. Second, some factors are both flows *and* funds simultaneously, regardless of T. To take Georgesçu's typical example, consider the production of hammers. Since hammers are produced (output), we need an $E_i(t)$ for hammers. Yet the factory might also use hammers in production and, if T is "small," we need an $S(t)$ to represent the hammer fund.

The flow/fund distinction illuminates an important difference between the employment of fund factors and flow factors: funds, such as a worker, are capable only of "binary" participation in production. One cannot send half one's body to work, although one can "goof off." The idleness of fund factors, particularly the capital fund, is seen by Georgesçu as a vital characteristic of factory production, and the elimination of fund idleness is seen as the fundamental source of the efficiency of the factory model.

To explain the role of fund idleness in factory production, consider any "elementary" process, that is, a sequence of actions using funds and flows, over a time interval [0, T], that produces 1 unit of output. Obviously, a process capable of producing 1 unit of output over [0, T] may well involve substantial idleness of some of the funds, particularly capital. Such a process can be thought of as a sequence (in a definite order!) of several "partial" processes. Now, suppose one wishes to produce more than 1 unit of output. There are (apparently) two ways in which this can be done. First, one could produce in parallel, called by Georgesçu the "external addition" of processes. Such an approach would yield no efficiencies. Alternately, one could produce in a line by staggering the elementary processes. This second approach, which may yield economic savings (by, among other things, reducing fund factor idleness) is called the factory system.

Georgesçu establishes the potential benefits of the factory system of production by proving a simple theorem on the sequencing of elementary (one output unit) processes.[23] In particular, if the output desired is sufficiently large, and if the variable periods of fund services are evenly divisible into the time of production, then there must exist a minimum number of elementary processes (level of output) such that the processes can be sequentially arranged so that every fund factor is continuously employed.

Thus, at least when output levels Q are large, the factory mode of production is a "steady state" of arbitrary duration with elementary processes starting continuously. The constituent flows are linear in t. Further, we must introduce the notions of "goods in process", called by Georgesçu the "process fund." Similar logic should apply to inventories. This leads to a simplification of our earlier representation of the production process to the equivalent forms $q = F(e; s)$ or $tq = \theta(te; ts)$ where θ is homogenous of degree +1.

Georgesçu goes further and introduces a large and rather confusing array of auxiliary relationships among the production flows, which are of two main sorts. First, maximal output rates q^* are regarded as constraints imposed by the "factory proper," that is, by the capital fund and labor fund: $q^* = G(K,L)$. Further, the first law of thermodynamics imposes materials balance constraints among the waste and maintenance flows.

What are the practical consequences of Georgesçu's formulation of the production process? There are several important implications for both theoretical and empirical use of production models. First, the logic of the factory system is clearly independent of the notion of "technology," but not of the magnitude of demand. Any processes which may be started at arbitrary times are liable to factory organization. Yet, as there are greater and greater numbers of "subtasks" involved in making a unit of output, the greater will be the number of "tasks" (output quantities) necessary to assure all fund factors are continuously utilized. Thus specialization is limited by the size of the market, and demands determine, to some extent, the technological recipes chosen.

Second, the requirement that tasks be able to commence at any arbitrary time is critical to obtaining the advantages of fully employed funds in production. Thus, we discover a critical difference between agriculture and industry. In farming, nature, not Man, ordinarily determines the sequencing of tasks. Fields must be harvested in parallel, not in a line. Only in cases such as poultry have we seen the successful application of the factory mode to agriculture. Thus, for at least the foreseeable future, we expect that agriculture will be characterized by high levels of fund factor unemployment (of an involuntary sort!), what amounts to "overcapitalization," and a persistent need for development of cottage industry of some viable sort.

Georgesçu's model of production also illustrates an important deficiency in the neoclassical notion of "substitution among factors." Can "capital" and "labor" be substituted one for the other? Presumably not in general, since there is no guarantee that, given any process, there must exist another process with, say, more capital (of the same kinds) and less labor (of the same kinds). To make all capital and labor funds homogeneous, cardinal variables, they are ordinarily measured in *dollars*, yet such an approach makes a mockery of marginal cost pricing theory, neoclassical cost minimization, and the neoclassical theory of production.

Perhaps the most notable difference between Georgesçu's approach and that of neoclassical production theory concerns the role of time in production modeling. Time, as Georgesçu argues, is an essential and, properly, explicit variable in any production model. To focus on the essentials, we consider the simpler conceptualization analyzed by Klein (1980). We have $q(t) = f(K(t), H(t), X(t))$ where $q(t)$ is the maximum rate of output crossing the boundary at t, $K(t)$ and $H(t)$ are the capital and labor funds present at t, and $X(t)$ are other inputs crossing the boundary at t. Then total output for the time interval $[0,T]$ is just

$$Q = \int_0^T f(K(t), H(t), X(t))dt$$

or, with constant rates of service, $Q = Tf(K, H, X)$, where Q is total output over the period. The important point is that the firm *selects* K, H, X, *and* T. The funds K and H are *amounts* of fund inputs, while X is a flow *rate* (per unit time). Note further that if $T = 0$, $Q = 0$: funds can be exhausted only over time.

The formulation outlined by Klein (1980) is perhaps the simplest that captures most of Georgesçu's main points. This template can be examined to evaluate time utilization via shift work and the resulting structure of cost. Typically, the capital *fund* must be irreversibly "bought" by the firm, while labor H and other inputs X can be purchased in a spot fashion. This causes a violation in the usual neoclassical symmetry of compensated demand cross price effects and results in a modified duality for Georgesçu-type flow-fund models.[24]

4.4 GEORGESÇU AND THE DEVELOPMENT OF AGRARIAN ECONOMIES

Due perhaps in equal measure to his Romanian background and his long association with the Graduate Program in Economic Development (GPED) at Vanderbilt University, Georgesçu analyzed and wrote about the economics of "peasant" societies throughout his professional career. As was typical of him, his interests focused not solely on "analytic similes" (models) of peasant behavior, but also on the historical and, indeed, geographic natures of those forces that created and, by long evolutionary process, changed the peasant societies with which he was familiar. Thus, in Georgesçu's writings on the food and population problems of agrarian societies, one finds a sustained focus on historical and sociological factors *augmented by* (rather than supplanted by, or supplanting) rigorous analytic work utilizing mathematical models of moderate to high complexity.

It is not merely the case, however, that Georgesçu discusses or recounts various sociological facts about peasant institutions, and then proceeds to a predictable, neoclassical analysis of the economic problems of production or employment. Such a "cobbling together" is wholly inconsistent with Georgesçu-Roegen's epistemological orientation. Rather, he was more interested in constructing models that were relevant to a particular economic environment, and this implied that the *assumptions* used were the critical issue, not the mode of analysis. Further, these assumptions must be *factually correct*. Methodologically, this meant that the role of assumptions on such data as the nature of production, substitution between inputs, flows and funds, and so on was more prominent than that of a large set of (often unstated) assumptions by Georgesçu on information, equilibrating processes, and the ability of agents to contract with one another. This approach is a characteristic reason why modern economists, trained in game theory and the like, often find his papers so confusing. Ordinarily, Georgesçu assumes competitive markets with perfect information, no search, price-taking behavior, and uniform price transactions only. This approach was so conventional during Georgesçu's "coming of age" as an economist that he continued to eschew stating it throughout his publishing career. Rather, he combined this straightforward methodology with such novel assumptions on technology and behavior as minimal consumption requirements (associated with biological survival). Thus, ordinarily we are given conventional models with very unconventional assumptions.

The combination of widely used analytic devices with very novel assumptions is perhaps more evident in Georgesçu's writings on agrarian development than anywhere else. To understand his still-controversial views on peasant economies, it is important to begin with a review of his sociological interpretation of peasant societies, their histories, and their material circumstances. Only by such a study can one appreciate the assumptions ("facts") Georgesçu uses for the construction of his analyses.

To begin, Georgesçu views peasant economies as fundamentally different from the urban/bourgeois economies treated by neoclassical and Marxist economists.[25] He feels that as a result, when these "orthodox" economists try to analyze peasant societies, they theorize in a vacuum, with little understanding, and even less sympathy, for such societies. Marx's contempt for the countryside is too well known to require re-elaboration here. Less apparent, but having equal effect, is the characteristic disregard by neoclassical economists for the material circumstances of the great bulk of mankind. For example, as Georgesçu was fond of pointing out, the Arrow–Debreu general equilibrium model assumes that every member of society (agent in the economy) is *endowed* with resources sufficient to maintain its life. Thus, the Arrow–Debreu-type general equilibrium model, which one might ordinarily criticize for its informational underpinnings, is actually guilty of the worse sin of irrelevancy as far as agrarian society is concerned. Neoclassical general equilibrium is a model of trade or exchange in some hypothetical "land of plenty." The economic problem of much of mankind – mere survival – is assumed away.

We therefore reach Georgesçu's first observation: a peasant economy is *not* the same as a farm economy, the latter being characterized by cash transactions, the profit motive, and so on. Additionally, and in a parallel fashion, agrarian economics is not the same thing as agricultural economics as understood by mainstream theorists.

Unlike the town, the peasant village is an organic form with a communal orientation. Like an organism, the continued existence of a peasant village requires a set of varied environmental factors, such as woodland, water, land suitable for pasture, and so on. Additionally, peasant villages are a scale-dependent phenomenon: they have optimal sizes and cannot exist successfully if they are too small or too large. Further, peasant society undergoes *evolutionary* developments, although, because societies may borrow or steal ideas from outside, evolutionary laws apply less to social forms than to their biological counterparts.[26]

The "oneness" of the peasant village is more, to Georgesçu, than an inoperative sociological label. The cooperation needed to support village institutions profoundly affects the determination of labor participation and compensation in agricultural production. Although everyone cares, to some degree, about others in their community, only villages are small enough so that these effects are not eliminated by "price-taking" type behavior. Georgesçu, in contrast to received opinion, held that neither blood kinship nor communality arising from working common land could credibly explain the cooperative character of village life, and he marshaled various arguments challenging these common explanations.[27] Rather, it was cooperation as a *Darwinian fitness* that was mostly responsible. This claim should probably not be taken literally: Georgesçu was referring to fitness with respect to *social* evolution. Yet such cooperation cannot be a fitness trait unless village size is less than a certain upper bound.[28] Additionally, like a gene, a village is (mostly) a stable structure resistant to rapid mutation. Since very rapid mutation defeats the advantages obtainable from random varietal introduction by foreclosing sufficient time to test the innovation in use, genes mutate only at a certain rate. Villages receive their stability from traditions, a resistance to rapid changes in practices that serves the same purpose.

The technological conditions of organic agriculture have shaped the form of village institutions to a great degree. In the first place, conflicts in such a system arise over flow rather than fund factors. The relevant problem is not "Who *owns* the land?", but rather "Who has claim to the outputs of the land?" The answer, on which Georgesçu felt almost all village institutions were based, is simple: only labor creates value, so labor must be the basis of division of the land's fruits. Thus, if one cleared land for farming, then one obtained a right to use it for a specified period. Yet one did not "own" it in the juridical sense of a transferable right to a fund, at least in the Romania of Georgesçu's youth. Thus, for Georgesçu, the common Austrian definition of property right/ownership – the "mixing" of one's labor with the land – is incorrect. By so "mixing", one obtained a finite claim to the output of the land created by one's labor, not a claim to the land itself. Georgesçu looked with a cynical eye on much of the literature on peasant economic development that emphasized land ownership as an index of social advancement.[29]

In addition to an enduring and, given the technological horizon, efficient attachment to a labor-based theory of income distribution, peasant societies were, in Georgesçu's experience, based on a strongly held commitment to an equal opportunity for all to work, though definitely not to any corresponding desire for equal incomes *ex post*. Everyone, in the beginning, had an equal right to work the land and retain that produce their labor made possible. Thus, unequal incomes could arise only due to differentials between the productivities of peasants' labor powers.

The large families observed in many peasant communities are themselves an indirect consequence of the egalitarian impulse. Redistributing land use periodically on a per capita basis rewards large families and, of course, evolutionary forces of a purely biological sort could be at work. Although these forces served peasant communities very well over millennia, technical conditions in agriculture, combined with the vanishing of unexploited woodlands in many areas, have resulted in dire population problems in many agrarian societies.

It is important at this stage to draw attention to a distinction in Georgesçu's writings which is frequently misunderstood. He believed that production in agriculture and production in manufactures differed fundamentally. The primary basis of this difference concerns the inability, for physical reasons, to (ever) achieve a factory system of production in most agricultural activities. Factories achieve their great efficiencies by sequentially running many elementary production processes in a staggered fashion so that the fund factors, such as capital and labor, are never idle. The assembly line, with its continuous flows of goods in process, is an obvious application. In agriculture, however, the sequencing and time spacing of activities such as plowing and harvesting are dictated by biology, and no factory organization is usually possible. This difference has a profound implication. In agriculture, undesired idleness of fund factors is unavoidable. This "unemployment" of factors is not desired because, if the worker could obtain a good crop by planting in December, he or she might well do so. December planting is, unfortunately, impossible in many cases.

The technical differences between agricultural and manufacturing technologies are, however, a different issue from the problems of agrarian societies, although of course these technical differences matter in the area of potential solutions. Georgesçu defined the agrarian problem as one of "overpopulation" (excess of labor) in a very specific sense. This definition, which we examine in detail below, allows one to talk about underdeveloped economies which are not, however, "overpopulated" in Georgesçu's sense. He felt that: (a) agriculture and industry differ in all societies, developed or not, overpopulated or not, due to fund unemployment phenomena imposed by nature; (b) societies may be underdeveloped or not; and (c) agrarian (peasant) societies may be overpopulated or not, although the overpopulated case is quite common and is an issue of enormous humanitarian concern. Georgesçu's most noted work concerns issues (a) and (c), while the problems of underdevelopment *per se* received a spottier treatment.

Let us turn first to the notion of "overpopulation" as used by Georgesçu in his analysis of the agrarian problem. What can "overpopulation" mean? The notion itself seems to imply there is some "optimal" population, an idea that appears beyond neoclassical economic analysis. Georgesçu's work on the Romanian national accounts, combined with other evidence, suggested that in some peasant societies large percentages of the rural able-bodied adult population could "disappear", yet agricultural output would not decline. For example, he cited evidence that, in 1914, Tsarist Russia maintained agricultural output although about 40 per cent of the male rural workforce went into the army at the outset of the First World War.[30] Whatever credibility one might assign to these historical claims, we do obtain a definition of overpopulation: a society is overpopulated if the marginal product of (unskilled) labor is zero.

Several objections to this formulation arise at once. First, it is admittedly difficult to tell the difference, particularly in the historical contexts discussed by Georgesçu, between a zero marginal factor productivity and slack. Workers might, for instance, labor only sufficiently to achieve some target level of income at least equal to subsistence. This explanation, however, could also violate neo-classical formulae, and thus explains one anomaly by invoking another.

Alternately, the circumstances cited by Georgesçu often involve significant social or political upheavals, and such environmental changes may change incentives and relative prices in significant ways. Whether output actually "stays the same" in some relevant sense is a key problem. Could output "stay the same" in moving from one marginally determined allocation of labor to another? Thus, the veracity of the historical examples given by Georgesçu, primarily using data from Russia (1914) and Romania (various years), is relevant. Georgesçu served as the chief statistician of Romania and was intimately involved with the compilation and classification of some of this data, so his opinion must carry considerable weight.

What are the "practical" implications of Georgesçu's formulation? While he offered several relatively complex models of agrarian economies, most of the substance of his argument can be obtained from a simple Edgeworth-box argument. We note first that, if labor's marginal productivity is indeed zero in agriculture, then efficiency must require the same result in the "other" sector, say manufactures. But then marginal pricing is inapplicable to this economy and cannot be used to allocate resources. Rather, the "feudal rule" (use all resources) must/will replace the "capitalist rule" of marginal productivity pricing. If one sector of the economy, such as manufactures, is organized on capitalist lines (so that relative wages equal the relevant price ratio), while zero marginal labor factor productivity exists in the countryside, then a very difficult problem of transitional planning arises and, according to Georgesçu, such a circumstance cannot be expected to self-correct. In general, economies cannot "jump" to marginal pricing regimes until sufficient capital accumulation occurs to raise the marginal factor productivity in agriculture to a level consistent with ongoing success. In Georgesçu's words: "no phase of economic development can be by-passed . . . feudalism cannot disappear before it has completely finished its job."[31] Indeed, the very low factor productivity of labor in agrarian economies leads to the perverse development of capital-intensive industries: in an overpopulated agrarian economy, wage/interest rate $> MP_L/MP_K$ for the marginal products of labor (L) and capital (K) since MP_L is "zero."

The agrarian problem has many political implications, the most notable being a serious conflict of interests between the town and the countryside. The urban cry for "cheap bread" is an obvious example. History also is replete, according to Georgesçu, with cases of advice from the "townees" disrupting the countryside and harming the peasants. Property rights themselves, in their juridical, transferable sense, are, in the case of land, inapplicable to the over-populated agrarian economy since such rights arise, in Georgesçu's view, from a basic conflict of interest between parties. Given the basic principles of the village, conflicts of this kind are not present. In a village, merit determines earnings, but *not* who gets to work. These institutional differences, and the lack of understanding of agrarian societies by mainstream neoclassical and Marxist economists, have poisoned communications and cooperation between town and countryside.

If the central factual claim of Georgesçu's analysis is correct, so that the marginal factor productivity of unskilled labor is zero in some agrarian societies, it is apparent that neoclassical prescriptions really may be useless. It is not possible to implement marginal-productivity based prices for labor in these circumstances. More generally, even in societies where labor's marginal product is positive, compensation based on marginal product pricing may not yield a subsistence wage. Yet, if neoclassicals are honest, they (we) must admit that the efficiency properties attributed to competitive pricing arise in models for which *the basic economic problem*, that is, survival, is not even an issue. Georgesçu's personal experiences and background made it impossible for him to ignore this irony.

4.5   GEORGESÇU, MARXISM, AND CAPITALIST BREAKDOWN

Georgesçu-Roegen became an economist in the 1930s and, like so many economists of that period, his "training" consisted of a far more extensive intro-duction to Marx and Marxism than is ordinarily the case today. Important debates on the interpretation of Marxist theory, particularly its mathematical representation, raged throughout the 1930s and 1940s, exemplified by the work of Joan Robinson, Paul Sweezy, Evsey Domar, and many other prominent analysts of the period. Although much of this literature has disappeared from graduate training in our times, Georgesçu consistently held that these debates provided valuable insights at any and all times. In the 1960s, when Georgesçu had considerable control over the content of PhD training at Vanderbilt University, Joan Robinson's writings on Marx remained required reading in the microeconomics core sequence long after other American graduate programs had dropped them.

Although Georgesçu was harshly critical of actual communist regimes and found many faults in Marx, his criticisms of Marxism strike the contemporary reader as peculiar in both emphasis and execution. Accustomed as we are to indictments of Marxism rooted in arguments of a humanistic character, Georgesçu's complaint that Marx committed errors such as adding inhomogeneous terms, so that a flow fuels its own growth through time, is likely to appear to us akin to the condemned man's complaint that the hangman's rope is too rough. Yet this incongruity is itself the source of an important insight into Georgesçu's character and into the great solemnity he attached to fundamental theoretical arguments in economics. He knew that if an economic theory (such as Marx's "simple reproduction") was logically incoherent, the game was over and no amount of additional argument could change this outcome. Thus, Georgesçu regarded the theoretical bases of economic models, whether Arrow–Debreu or Marx–Sweezy, with very great seriousness. The fundamentals always mattered more than anything else.

Georgesçu's published work and private conversations frequently made reference to Marx, although he wrote relatively few works explicitly dealing with Marxism. An important exception to this rule is his famous paper, "Mathematical proofs of the breakdown of capitalism," published in *Econometrica* in 1960. This paper is important for several different reasons. First, it gives the reader the best illustration available of Georgesçu's interpretation of Marxist economics. The now commonplace distinction between "Marx and the Marxists" is operative here. Second, the paper represents an important, independent contribution to economic theory. Georgesçu offers a coherent, analytically correct version of the so-called Bauer–Sweezy model of the failure of capitalist accumulation. Several errors made by Sweezy in his 1942 book, *The Theory of Capitalist Development*, are corrected by Georgesçu. Having honestly made the best analytic case he could for Marx's conclusion, Georgesçu offers a very careful analysis and concludes that Marx's "forecasts" of the output of consumer goods exceeding demand (causing a "breakdown") are invalid and arise from (rather subtle) technical errors. Because of the importance of these points, we briefly review the Sweezy arguments and Georgesçu's refutation of them.

Georgesçu begins by stating a candidate mathematization of Marx's model of capitalist accumulation. The flow of net national income y is divided between a consumption flow c and net accumulation a. Some of y forms a wage flow w to the working class, while some is surplus value s, so $y = w + s$. Further, $s = \ell + a$ where ℓ is capitalist consumption. Accumulation a is in two parts, increments to variable capital v (with stock V), and increments k to constant capital K. (Variable capital is subsistence goods for workers, constant capital is capital in the usual sense.)

Next, the common Marxist assumptions of proportionality between the constant capital stock K and consumer good flows (K $= \lambda(w + l + v)$), and between variable capital and wages (V $= \mu\, w$), are made. As Georgesçu points out, the constants λ and μ are *dimensional* – that is, measured in time units. Thus, we can take, say, $\mu = 1$ so that V $= w$. Note also that there is no relationship between labor services inputs and output since, with Marx, labor usage is purely a decision of the capitalist class.

Adopting the ordinary Marxist assumptions on capitalist behavior implies that (a) $a = a(s)$, $\partial a\,/\,\partial s > 0$ and $(\partial\,/\,\partial s)\,(a\,/\,s) > 0$; (b) $k = k(a)$, $\partial k\,/\,\partial a > 0$, $(\partial\,/\,\partial a)\,(k\,/\,a) > 0$. This means that surplus value is accumulated in greater proportion as surplus increases, and that, as accumulation rises, greater proportions of accumulation are invested. These assumptions were believed by Sweezy to guarantee "breakdown" of the capitalist system.

We finally add the dynamic relation $\partial V\,/\,\partial t = v$ and $\partial K\,/\,\partial t = k$. With this, we have a (fairly complicated) dynamic system which Georgesçu offers as a sensible representation of the Marxist template. He points out that his formulation differs from that of Sweezy in two ways. First, Sweezy takes surplus value as being composed of the terms $s = \ell + v + k + (\partial\ell\,/\,\partial t)$ as opposed to Georgesçu's $s = \ell + v + k$. The inclusion of $\partial\ell\,/\,\partial t$, which reflects changes in capitalist consumption through time, was a source of controversy among Marxists who lacked the mathematical sophistication to see that since $\partial\ell\,/\,\partial t$ is of a different dimension than ℓ, they cannot be added together. Moreover, Georgesçu felt that this sin reflects a deeper (the deepest) technical Marxist fallacy, which holds that "a material flow can be the source of its own growth."[32] A similar error, with the same epistemological interpretation, occurs in the expanded expressions for national income given by Georgesçu and Sweezy.

Georgesçu succeeded in formulating a "corrected" version of the Sweezy–Bauer mathematical representation of the Marxist economy. He was then able to evaluate the general Marxist forecast of inadequate demand for consumption goods or "overproduction."

Georgesçu characterizes Sweezy's (conventional) position as follows. First, if $\partial y\,/\,\partial t > 0$ (national income is growing), then $\partial k\,/\,\partial t > 0$ (increments to capital are growing). Second, if $\partial^2 y\,/\,\partial t^2 < 0$ (the growth is slowing), then $\partial k\,/\,\partial t < 0$ (increments to capital must be declining). Yet, together, if, $\partial^2 y\,/\,\partial t^2 < 0$, then

$\partial k / \partial t$ must exceed its *equilibrium* value, so that output of consumption goods exceeds demand.

Georgesçu demonstrates that the Sweezy argument is invalid because its first and second postulates are wrong. As is usually the case in such tasks, the basis of his proof is by counterexample. Georgesçu finds a solution (to the system of equations representing the dimensionally corrected Sweezy–Bauer analysis) that violate Sweezy's claims. In fact, there are solutions corresponding to "strongly growing" (that is, v and ℓ are accelerating) capitalist economies such that $\partial^2 y / \partial t^2 < 0$.

We thus reach the following conclusions. Mathematical representations of Marx's capitalist economy typically suffer from dimensional inhomogeneity problems. This, in turn, implies that Marx's forecast of overproduction must be evaluated in a model that corrects these mathematical errors. Having done this, Georgesçu concludes that Marxian predictions are not borne out by the mathematics. In other words, those stylized facts that one would associate with Marxist positions do not, by themselves (even when true), imply a breakdown as some believe.

Despite Georgesçu's fruitful analysis, he saw in the problem of modeling capitalist breakdown a kind of paradox: the problem is really, in principle, impossible. After quoting Schrodinger on the limitations of mathematics, Georgesçu remarks:

> capitalism, like all other economic systems that preceded it and that will be produced by the continuous evolution of human society, is a form of life. Some aspects of its functioning lend themselves perfectly to mathematical analysis. Yet, when we come to the problem of its evolution, of its mutation into another form, mathematics proves to be too rigid and hence too simple a tool for handling it.[33]

4.6 CONCLUSION

Nicholas Georgesçu-Roegen made many original and important contributions to standard economics. At least two of these, his early work on integrability and on Leontief models, could be considered very important foundational results. In addition, even when he analyzed "conventional" types of problems, he often proposed novel solutions that were, in many cases, far ahead of their times. For example, his work on hysteresis predates Becker's important application of consumption theory to "Rational Addiction." Ironically, those who knew Georgesçu-Roegen unanimously speculate that he would not have agreed with the use of standard economics to treat such issues.

Although in many senses a pioneer, Georgesçu-Roegen was clearly, in conventional economics, a scholar of the mid-twentieth century. When he died, the *New York Times* eulogized him as a "grand old man" of economics. This is a fair assessment both in the senses that Georgesçu knew such scholars as Schumpeter and that his work reflected the main interests of those times. Absent from Georgesçu's opus is anything explicitly connected with such "modern" topics as incentives, game theory, principal–agent problems, information, and so on, although as noted he was intuitively familiar with some of these ideas.

Inevitably, a number of very worthwhile ideas in Georgesçu's work have been omitted in this too brief review of his writings. In fairness, one should mention at least three of these. First, Georgesçu produced some interesting early work on the "constancy" of the marginal utility of income.[34] Although the modern reader will associate this idea with welfare calculation and quasi-linear utility functions, Georgesçu saw the problem as deeper than that. In fact, he expressed the view that the constancy of the marginal utility of income was important in making choice theory less "static," so that consumers could be envisioned as not deciding on the entire consumption profile at one instant, but instead sequentially making purchases through time.

Second, Georgesçu proposed the application of relaxation phenomena to economic models. Such phenomena are important in physics, and refer to periodic phenomena exhibiting two distinct phases of change, representing regime shifts. The great physicist van der Pol introduced this concept to physics in 1926, and attempted to represent both phases of motion by a single equation. Georgesçu applied the idea to business cycles and macroeconomic policy, and introduced the concept of "phase periodicity" to describe such systems.[35]

Finally, Georgesçu made an important theoretical contribution to the notion of "complementary goods" in his essay, "A diagramatic analysis of complementarity" (1952). Goods which, when consumed, strongly affect the utility derived from other goods have long interested economists, with early work going back at least to Pareto and Edgeworth. Georgesçu shed considerable light on the interrelations between definitions based as utility changes and those based on money/demand data. Further, he proposed "isotimetic loci," sets composed of bundles selected so that some good exhibits constant marginal utility, to study complementarity problems. The article even includes an early analysis of separability in utility functions from the functional equations perspective. The issues raised by Georgesçu in this article may be quite important in the evaluation of environmental amenities, and interest in this work is increasing.[36]

Georgesçu-Roegen's work within the confines of conventional economics is, by itself, sufficient to assure him a place as an important modern economist. Yet his legacy will not be built on these foundation stones, though they are sturdy. Posterity will undoubtedly judge Georgesçu in light of his contribution, judged retrospectively, to environmental and resource economics. It is to this topic that we now turn.

NOTES

1. See, for example, Hurwicz or Uzawa in Chipman *et al.* (1971).
2. See N. Georgesçu-Roegen, "The Pure theory of consumers' behavior" (1936) for several examples.
3. Indeed, in line with familiar Austrian criticisms, Georgesçu felt that there really was no "choice" in neoclassical choice theory. This belief kept alive his interest in issues surrounding the "constancy of the marginal utility of income," which he felt could be used to "temporally disaggregate" choice problems. Again, this idea is based on introspection.
4. Varian (1992: 125).
5. For WARP, see Samuelson, *Foundations of Economic Analysis*, 1983, ch. 5. The term WARP first appears in Samuelson (1950a). SARP is due to Houthakker (1950). Varian (1992) provides a summary.
6. This point is still abused.
7. Georgesçu met R.G.D. Allen and later stated that he believed his chilly reception arose from his discovery of Allen's error.
8. This phenomenon arises, of course, because of the necessity of making certain rather restrictive assumptions to define "non-binary" preferences. We discuss this below.
9. Micro changes are hysteresis, while macro changes of irreversible character arise from entropic degradation, scientific progress, and social evolution.
10. See, for example, Georgesçu-Roegen in Szenberg (1992).
11. Becker's idea, of course, is not identical in spirit to Georgesçu's.
12. For example, if the homeowner incurs costs from effort and fatigue in lawn watering, then one should not claim that water has differing marginal utilities per dollar based solely on the fact that, if it were effortless, he or she would water the lawn more.
13. *Given* the other axioms of expected utility theory, Georgesçu often commented that the independence axiom was "perfectly sensible." Given his methodological outlook, this is odd.
14. Obviously, though, people are seen to trade money for lottery tickets, stocks, and other uncertain instruments. Georgesçu would probably respond that circumstances are always risky to begin with.
15. Thus, there is no "marginal rate of substitution" between bias and variance as, for example, with minimum MSE approaches.
16. N. Georgesçu-Roegen's, *The Entropy Law and the Economic Process* (1971) offers a detailed analysis.
17. The Austrian concept of "roundaboutness" of production, and emphasis on the period of production and intermediate goods, is much in the spirit of Georgesçu's approach. Faber *et al.* (1987/1995) offer a very interesting treatment.
18. This issue is closely related to Georgesçu's oft-expressed view that there is an important distinction between desired leisure, undesired leisure, and the disutility of work.
19. This incident is recounted in Georgesçu's contribution to Szenberg (1992).
20. Georgesçu-Roegen's *The Entropy Law*, p.222.
21. Ibid., p.227. This idea appears worth investigation.
22. Thus, rested men go into work and tired men come out.
23. See N. Georgesçu-Roegen's "Process in farming versus process in manufacturing: a problem of balanced development" (1965) for the proof.
24. See Klein (1980) for further details.
25. The reader can find multiple statements of these views in *The Entropy Law*, "Economic theory and agrarian economics" (1960), "Process in farming versus process in manufacturing", and "The institutional aspects of peasant communities: an analytical view" (1965).
26. There is a recurring tension in Georgesçu's work. On the one hand, he stated many times that societies evolve, and that this process cannot be understood as solely a consequence of utility-maximizing behavior by agents. Thus, institutions are *not* just contracts. Neither, however, are they pure accidents: they respond to pseudo-Darwinian fitness influences. This is closely related to Georgesçu's observation in the physical context that there are many properties of

compounds (such as water) which can never be deduced from study solely of the properties of their constituent parts.

27. See Georgesçu-Roegen, "The institutional aspects of peasant communities" (1969).
28. The optimal scale of the village arises, according to Georgesçu, from primarily technical constraints on agricultural production. He compared these constraints to physical ones limiting the sizes of organisms.
29. Because of the important technological differences between production in agriculture and manufactures, village land "ownership" conventions differ from property rights in the towns.
30. Georgesçu relied on Hubbard (1939) for some of his data, but he saw this at first hand in Romania.
31. Georgesçu-Roegen, "Economic theory and agrarian economics" (1960: 35).
32. N. Georgesçu-Roegen, "Mathematical proofs of the breakdown of capitalism" (1960: 231).
33. Ibid., 243.
34. See N. Georgesçu-Roegen, "Marginal utility of money and elasticities of demand" (1936), and "Revisiting Marshall's constancy of marginal utility of money" (1968).
35. N. Georgesçu-Roegen, "Relaxation phenomena in linear dynamic models", in Koopmans (1951).
36. We are indebted to Professor V. Kerry Smith for pointing out the current interest in this work.

5. An economist's primer on thermodynamics

5.1 INTRODUCTION

Questions about exhaustible resource economics invariably involve the future, for when but in the future could exhaustion of resources occur? Since economists cannot have access to data about the future, some understanding of scientific principles is necessary to avoid making unwarranted assumptions about future technological conditions. For example, few economists would want to model the future of the airline industry assuming any strictly positive probability that an anti-gravity paint will be discovered. What makes such an analysis uninteresting to economists is their familiarity with the law of gravity. Learning about the second law of thermodynamics (the "Entropy Law") serves a similarly useful function.

Nicholas Georgesçu-Roegen felt deeply that economic theory should accurately reflect the concrete realities of economic life. He intended nothing the least mystical or metaphysical when he named his *magnum opus The Entropy Law and the Economic Process* (1971). His intention was for his view of entropy to be the same as that of an applied chemical or metallurgical engineer, and that is the view we wish to explain in this chapter.

5.2 THE LAWS OF THERMODYNAMICS

The First Law of Thermodynamics states that in an isolated system, energy is neither created nor destroyed. The First Law is true (barring exchanges between matter and energy according to Einstein's $E = mc^2$, where E stands for energy, m for matter, and c for the speed of light). However, this First Law cannot explain why an ice cube placed on a sidewalk on a hot summer day always melts. The First Law would permit heat to flow from the ice cube as long as an equal amount of heat flowed to the air – yet this would make the ice cube colder and the air warmer, which never happens.

Entropy is the concept scientists use to explain why ice cubes on a hot sidewalk melt, and in general why heat always flows from hotter objects to colder ones and why spontaneous physical processes are spontaneous. Let T stand for temperature in degrees Kelvin, denoted °K. Kelvin temperature is 273.15° higher than Celsius temperature. Let Q stand for the flow of heat into a material (or system, which is a collection of materials). The metric unit of heat Q, as well as that of work, is the Joule, denoted J; the English unit is the calorie or the BTU. Then the change in the entropy S of the material (or system) is defined to be

$$dS = dQ_{rev}/T, \tag{5.1}$$

where the subscript "rev" stands for a reversible process; physicists call a process "reversible" if it involves no dissipative effects such as friction, viscosity, inelasticity, electrical resistance, or magnetic hysteresis (Zemansky, 1968: 193, 215; Mackowiak, 1965: 59). As Zemansky (ibid.: 225) adds; if a system undergoes an irreversible process between an initial equilibrium state i and a final equilibrium state f, the entropy change of the system is equal to the integral from i to f of dQ/T, taken over any reversible path from i to f. No integral is taken over the original irreversible path.

Suppose one has an isolated system containing two bodies, one hot and one cold, placed in thermal contact. When a given quantity of heat, Q_0, flows from the hotter body to the colder one, the change in entropy of the system is:

$$\Delta S = \frac{-Q_0}{T_{hot}} + \frac{+Q_0}{T_{cold}}.$$

Since $T_{hot} > T_{cold}$, $\Delta S > 0$. If heat were to flow in the opposite direction, away from the colder body and toward the warmer body, then the Q_0 terms in the above equation would change sign and ΔS would be negative. Because it is the former and not the latter which is always observed, nineteenth-century physicists postulated that in an isolated system entropy never decreases. This is the Second Law of Thermodynamics.

Notice that (5.1) defines only changes in entropy; absolute amounts of entropy are not defined. In Georgesçu-Roegen's terminology (*The Entropy Law and the Economic Process*, 1971: 100, 146), variables which have this property are said to be "weakly cardinal." Formally, a variable x is weakly cardinal if and only if no physical predictions will change when x is replaced everywhere with $x + C$ where C is a constant. Other examples of weakly cardinal variables are gravitational and electromagnetic potential.

With this background, a systematic list of the laws of thermodynamics follows.

Zeroth Law   Two systems which are in thermal equilibrium with a third are in thermal equilibrium with each other (Dugdale, 1996: 13).

First Law   During a process in which no heat is exchanged with the environment, the work done is only a function of the initial and final states of the system, not of the path. In addition, during any process, heat flow, Q, is equal to $U_f - U_i + W$ where U_f and U_i are the final and initial internal energies of the system and W is the work done by the system (ibid.: 20; Zemansky, 1968: 78–9).

Second Law, Kelvin–Planck statement   No process is possible whose *sole* result is the absorption of heat from a reservoir and the conversion of this heat into work (Zemansky, 1968: 178).

Second Law, Clausius statement   No process is possible whose *sole* result is the transfer of heat from a cooler to a hotter body (ibid.: 184).

Second Law, entropy statement   In an isolated system, entropy is non-decreasing (ibid.: 234).

Third Law, unattainability statement   It is impossible to reach absolute zero by any finite number of processes (ibid.: 498; Dugdale, 1996: 177).

Third Law, Nerst–Simon statement   In the limit as temperature goes to $0°K$, the entropy change of any reaction is zero (Zemansky, 1968: 498; Rao, 1985: 257; Dugdale, 1996: 160–1).

Proofs that the three statements of the Second Law are equivalent can be found in Zemansky's text, as can proof that the two statements of the Third Law are equivalent.

Much has been deduced from each of these statements. We have space to discuss only a few of these implications, and they will mostly involve chemical, not mechanical, engineering. However, the most famous implications for mechanical engineering should be mentioned. Steam engines and internal combustion engines are "heat engines" since they use temperature differences to do their work. (A car's motor will not work in a room whose temperature is above the flash point of gasoline.) Define the "efficiency" of a heat engine to be its work output divided by its heat input. Then it can be shown that the Second Law implies that the maximum efficiency of any heat engine operating between the temperatures T_{low} and T_{high} is $1 - (T_{low}/T_{high})$. This is less than 1 because T_{low} is strictly greater than zero due to the unattainability statement of the Third Law. The only type of heat engine which can attain maximum efficiency is a "Carnot engine," which is defined to be "a reversible engine operating between only two reservoirs." The cycle in which a Carnot engine operates is called the "Carnot cycle," which, when represented on a temperaturevs.-entropy diagram, is a rectangle. It will never be possible to construct a Carnot engine because Carnot engines are reversible, which real engines can never be (plagued as they are by friction, viscosity, turbulence, and the like). Georgesçu-Roegen summarized (*Energy and Economic Myths*, 1976: 11): "as we know from Carnot, in each particular situation *there is a theoretical limit independent of the state of the arts, which can never be attained in actuality.*"[1]

5.3   ENTROPY

One of the most surprising consequences of equation (5.1) comes from analyzing the mixing of two ideal gases in a container whose walls allow no heat to flow. (An "ideal" gas has molecules which do not interact with their neighbors.) At first sight, one might conclude from (5.1) that since no heat can flow into or out of the container, the system's entropy cannot change. This is incorrect. Entropy change is not $\int_i^f dQ/T$ (which is zero in this case) but rather $\int_i^f dQ/T$ *over a reversible path.* The mixing of two ideal gases is not reversible. The only way to calculate the entropy change accompanying the mixing of two ideal gases is to replace the original irreversible path linking the states before and after the mixing with a reversible path linking those same states, then performing the integration. Because entropy change is a function of the initial and final states only, entropy change along the new reversible path (which can be calculated using (5.1)) must be the same as the entropy change along the original irreversible path. Along the reversible path, heat does flow.

To derive the result, instead first consider a slightly different process:

> Imagine a thermally insulated vessel with rigid walls, divided into two compartments by a partition. Suppose that there is a gas in one compartment and that the other is empty. If the partition is removed, the gas will undergo what is known as a *free expansion* in which no work is done and no heat is transferred. From the first law, since both Q and W are zero, it follows that *the internal energy remains unchanged during a free expansion.* (Zemansky, 1968: 115)

When an ideal gas's internal energy does not change, its temperature does not change (ibid.: 5–7 or 291); it follows that the free expansion of an ideal gas is isothermal (has constant temperature). A free expansion is irreversible. As Zemansky writes (ibid.: 226–71):

> to calculate the entropy change of the system, the free expansion must be replaced by a reversible process that will take the gas from its original state (volume V_i, temperature T) to the final state (volume V_f, temperature T). Evidently, the most convenient reversible process is a reversible isothermal expansion at the temperature T from a volume V_i to the volume V_f.

Make this replacement; call the old "irreversible, free expansion, insulated vessel" path 1, and call the new "reversible isothermal non-insulated vessel" path 2. Since path 2 is isothermal, the internal energy U of the ideal gas along path 2 does not change (ibid.: 5–7 or 291). Since $Q = U_f - U_i + W$ according to the First Law, along path 2, dQ/T equals $(0 + dW)/T$. This will be strictly positive because when an ideal gas undergoes an isothermal expansion, heat must flow into the gas (unlike path 1, when the container's walls prevented heat flow). The work dW done on path 1 was zero, but on path 2 it is PdV where P is pressure and V is volume. It follows that on path 2 $dQ/T = P\ dV/T$ or, using the ideal gas law, $PV = nRT$ where n are the number of moles and R is the universal gas constant, that:

$$\frac{dQ}{T} = \frac{nRdV}{V}.$$

The equation (5.1) implies that the entropy change for path 2 (which is reversible) is:

$$S_f - S_i = \int_{T_i}^{T_f} \frac{dQ}{T} = \int_{V_i}^{V_f} \frac{nRdV}{V} = nR \ln \frac{V_f}{V_i} \qquad (5.2)$$

(ibid.: 227). While this was calculated along the reversible path 2, it must also be the entropy change for the irreversible free expansion of path 1, because the initial and final states (volume, temperature, and pressure) of the gas are the same on each path. (See also Dugdale, 1996: 60.)

Now consider the problem posed at the beginning of this section: a container whose walls allow no heat to flow is divided into two compartments, each containing a different ideal gas, and then the wall separating the two gases is punctured. As the gases mix, the entropy of each gas will increase according to (5.2), because the diffusion of two ideal gases is equivalent to two separate free expansions (Zemansky, 1968: 228, 562). Suppose the first ideal gas was at pressure P before the mixing occurs; then $PV_{i1} = n_1RT$. Suppose the second ideal gas was at pressure P before the mixing occurs; then $PV_{i2} = n_2RT$. The sum of V_{i1} and V_{i2} is the total volume of the container; if we call this V then $V_{f1} = V = V_{f2}$ and $PV = (n_1 + n_2)RT$. From the previous two sentences it follows that the entropy change occurring when the two gases mix is:

$$\left(S_f - S_i\right)_1 + \left(S_f - S_i\right)_2 = n_1 R \ln \frac{V_{f1}}{V_{i1}} + n_2 R \ln \frac{V_{f2}}{V_{i2}}$$

$$= n_1 R \ln \frac{n_1 + n_2}{n_1} + n_2 R \ln \frac{n_1 + n_2}{n_2}$$

$$= R \left[n_1 \ln \frac{n_1}{n_1 + n_2} + n_2 \ln \frac{n_2}{n_1 + n_2} \right]. \qquad (5.3)$$

The entropy of the system is the sum of the entropy of its two components, and therefore the entropy rise of the system equals (5.3). This entropy change can be detected experimentally using the calorimetric technique described in section 5.5.

The remarkable result that mixing two gases increases entropy has led observers who do not understand that (5.3) comes from (5.1) to think that there is more to the idea of entropy than expressed by (5.1) – that entropy measures increasing mixing, or spatial disorder, or even some general notion of "order" *per se*. This is not a scientifically valid stretch, and we will always define entropy by (5.1) alone.

On the other hand, scientists do sometimes find it useful to distinguish the contribution to entropy change represented by (5.3) from other contributions to entropy change. When this is done, the component of the entropy change given by (5.3) is called "configurational entropy" or "the entropy of mixing," and the other components of the entropy change are together called 'thermal entropy." The sum of these two components equals total entropy change, which we will continue to call simply "entropy change," and which continues to be defined by (5.1). The choice of the term "thermal entropy" is unfortunate, because it can mislead one into thinking that configurational entropy has nothing to do with thermal changes, and hence that configurational entropy has nothing to do with (5.1). Configurational entropy is actually just as much a component of (5.1) as thermal entropy is.

Before leaving the topic of (5.3) it is important to note that while (5.3) is valid for ideal gases, it is usually only part of the story for other kinds of systems – the rest of the story being the so-called thermal entropy. For example, while the direction of increasing configurational entropy for a system of oil and water, or metallurgical slag and matte, is to mix, it turns out that the direction of increasing (total) entropy for a mixture of oil and water, or metallurgical slag and matte, is to separate, not to mix. We will discuss the reasons for this below; for now it suffices to keep in mind that physical mixing does not always increase entropy, and separation or purification does not always decrease it.

A related point can be made in regards to chemical, as opposed to physical, mixing. Even in everyday life there are many substances which do not spontaneously chemically combine when put next to each other. Most of these are instances in which increasing entropy favors chemical separation or purification, not mixing or chemical bonding.[2]

It must be pointed out that there are substances which are found in nature in a state which is not their state of highest entropy – that is, not their stable equilibrium state. The Entropy Law does not specify the rate of reaction, and the reactions which would bring these substances to thermodynamically stable equilibrium occur at a rate of zero. The reason is that such "metastable" systems have an "activation energy" barrier which blocks their path to their truly stable configuration. A catalyst works by supplying the activation energy, which is then returned in full to the catalyst, in an analogous way as the energy required to lift a boulder to the top of an embankment can be fully recovered by the time the boulder passed its original altitude on its way down the mountainside.[3]

Further insight into the entropy concept occurred after scientists began to appreciate that the statement "in an isolated system, entropy never decreases" involves time in a fundamental way. The statement more precisely means that "in an isolated system, as time goes forward, entropy never decreases." No other physical law distinguishes between time going forward or going backward. It is not possible to determine whether a film of a billiard ball rolling on a frictionless surface is showing the motion as it occurred or is showing it backwards. Newton's laws of motion, which govern the billiard ball's behavior, do not distinguish between time going forward or backward. Only the Second Law of Thermodynamics defines "time's arrow." In so doing, it explains the direction of all physically or chemically spontaneous processes.

While the so-called "classical thermodynamics" of the mid-nineteenth century, which we have been presenting, predicts phenomena correctly, it contains no explanation of *why* the entropy of, say, 10 grams of copper should equal its experimentally determined magnitude. Insight into such topics was accomplished by Ludwig Boltzmann in his development of statistical mechanics, which is in some sense a modern version of classical thermodynamics.

5.4   ENTROPY IN STATISTICAL MECHANICS

It is easiest to discuss statistical mechanics' "explanation" of (5.1) by treating thermal entropy and configurational entropy separately. In statistical mechanics, thermal entropy concerns the number of ways in which the non-identical or identical particles can be mixed or distributed over the available energy levels (Gaskell, 1981: 85), while configurational entropy concerns the number of ways in which non-identical particles "can be mixed or distributed over the available positions in space" (ibid.: 85, 88). The total entropy change is the change of thermal entropy plus the change of configurational entropy (ibid.; Wicken, 1987: 2–4).

Statistical mechanics' notion of configurational entropy is rather easy to understand. As an example, follow Rao (1985: §7.11) in calculating the entropy of a solid solution "such as solid copper-nickel alloy." Let N_1 be the number of copper atoms and N_2 be the number of nickel atoms. The number of ways in which N_1 indistinguishable particles of type 1 and N_2 indistinguishable particles of type 2 can be ordered is $(N_1 + N_2)!/(N_1! N_2!)$. This is the "thermodynamic probability" W for the copper and nickel atoms' spatial (or "configurational") distribution on the crystal lattice sites. The statistical mechanical formula for entropy is:

$$S = k \ln W + S_0, \tag{5.4}$$

where k is Boltzmann's constant (1.38×10^{-23} J/°K) and where the constant S_0 is, by tradition but not by necessity, set equal to zero. (An excellent derivation of (5.4) is given by Dugdale, 1996: 95–9.) Following tradition, by setting the constant equal to zero, (5.4) gives the entropy of the copper-nickel alloy as:

$$S = k \ln \frac{(N_1 + N_2)!}{N_1! N_2!}.$$

Applying Stirling's Approximation $\ln(x!) \approx x \ln x - x$ yields:

$$S = -k \left[N_1 \ln \frac{N_1}{N_1 + N_2} + N_2 \ln \frac{N_2}{N_1 + N_2} \right].$$

Since $k = R/N_A$ where R is the universal gas constant (8.31 J/(°K · mole) and N_A is Avogadro's Number (6.022×10^{23}/mole), one has:

$$S = -R \left[\frac{N_1}{N_A} \ln \frac{N_1}{N_1 + N_2} + \frac{N_2}{N_A} \ln \frac{N_2}{N_1 + N_2} \right].$$

The number of moles n_i is equal to N_i/N_A, so:

$$S = -R \left[n_1 \ln \frac{n_1}{n_1 + n_2} + n_2 \ln \frac{n_2}{n_1 + n_2} \right]. \tag{5.5}$$

This is identical to (5.3), though the derivations – one using statistical mechanics and the other using classical thermodynamics and applying to ideal gases – could not have been more different.

Rao (ibid.) continues with another example. This one shows how statistical mechanics can generate testable predictions. Suppose we wish to calculate the configurational entropy of solid carbon monoxide at $0°K$. There are 2^2 different ways of arranging two carbon monoxide molecules randomly in a row (CO CO, CO OC, OC CO, OC OC), and there are 2^N different ways of arranging N carbon monoxide molecules randomly in a row. Hence the configurational thermodynamic probability W of a system composed of N_A carbon monoxide molecules randomly arranged is 2^{N_A}, and the configurational entropy of such a system is $S = k \ln W = k \ln 2^{N_A} = k N_A \ln 2 = R \ln 2 = 5.8 J/°K$. On the other hand, the configurational entropy of a system of solid carbon monoxide at $0°K$ arranged, not randomly. but instead as a perfectly ordered crystal, would be $S = k \ln 1 = 0 J/°K$ because there is only one such perfect arrangement (namely CO CO CO CO CO ...). Experimental measurements have found the configurational plus thermal entropy of one mole of solid carbon monoxide at $0°K$ to be $4.2 J/°K$. Rao apparently assumes that the thermal entropy of this solid is zero, because the temperature is absolute zero; on this assumption, the configurational entropy of solid CO at $0°K$ is $4.2 J/°K$. Since 4.2 is between 0 and 5.8, Rao concludes that "the molecular distribution within the crystalline carbon monoxide is partially ordered (i.e., less than completely random)" (and less than perfectly ordered). Being able to predict that the entropy of solid CO at $0°K$ is between 0 and $5.8 J/°K$ is wholly outside the capabilities of classical thermodynamics.[4]

Statistical mechanics' notion of thermal entropy is not as easy to understand as its notion of configurational entropy. A hypothetical example will convey its basic idea. Suppose there are two electrons orbiting an atom and suppose there are three orbits (three energy levels) around the atom. Suppose you know the sum of the energy of the two electrons, and suppose this sum implies that the two electrons are either both together in level B, or one of them is in level A and the other in level C. You cannot observe which level each electron is in. Statistical mechanics predicts that one of the electrons is in level A and the other is in level C. It arrives at this prediction by borrowing from the mathematics of probability: there are two ways to have "one electron in A and one in C" while there is only one way to have "both electrons in B," so if states were to occur randomly, the first state would have probability 2/3 and the second state would only have probability 1/3. Entropy is defined as an increasing function of the "probability" of a configuration, so the law of increasing entropy becomes a prediction that systems move toward configurations of higher "probability."[5]

In non-ideal gases, liquids, and solids, the calculation of thermal entropy involves all kinds of energy present in the system. For example, each particle could have nuclear, electronic, translational, rotational, and vibrational energy. In addition, there could be intermolecular interactions, even spherically asymmetric ones if different parts of the molecules are different, as is the case with oil and water molecules.[6] Finally, the assumption of the particles occupying infinitessimal points in a regularly spaced lattice will be violated if the particles are of unequal size or are of non-negligible size compared to the intermolecular distances, as can happen with organic molecules. Many of these deviations from ideality involve the particles' spatial orientation, yet they are categorized as "thermal" rather than "configurational" entropy. This shows that there is artificiality in dividing entropy into these two categories; the thermal, or energetic, properties of matter often come from the configuration which the particles are in. Still, the distinction between the two components of entropy can be retained by remembering that configurational entropy has to do with the gross arrangement of the particles and nothing else.

While configurational entropy increases as substances mix, total entropy may fall when substances mix, once all these other factors from thermal entropy are taken into account. As we mentioned above (p. 88), this happens with oil and water and with metallurgical slag and matte. Physical mixing does not always increase entropy and physical separation does not always decrease it.

Examples of successful predictions of statistical mechanics are numerous. They include Einstein's 1907 prediction of the temperature dependence of a solid's heat capacity at constant volume; the prediction of an ideal gas's heat capacity at constant volume, both for high temperatures, and for low temperatures where quantum effects are important; and the prediction of the standard entropies of argon, oxygen, and nitrogen (Dugdale, 1996: 105ff, 121–2, 134, 159–62). ("Standard" entropies are defined with an initial condition of $0°K$ and one atmosphere pressure and a final condition of $300°K$ and one atmosphere pressure.) Physicists and chemists continue to extend the reach of statistical mechanics to more and more complicated systems, though analyzing a system as complicated as oil and water is still well beyond what textbooks present.

Since the Entropy Law of classical thermodynamics explains time's arrow for heat flow or for two previously separated gases mixing in a container, it was natural to wonder whether increasing entropy explained time's arrow for other phenomena, such as a raw egg splattering on a hard surface, the shuffling of a deck of cards, or even the development of complex social organizations. Statistical mechanics sheds no more light than classical thermodynamics does on these phenomena, because it does not throw out the basic definition of entropy, (5.1), but merely explains it with (5.4). Unless a phenomenon can be described by heat flow Q and temperature T, it has nothing to do with either the classical or the statistical theory of entropy and the Second Law of Thermodynamics.

5.5   ENTROPY IN ENGINEERING

In order to raise the temperature of water, it is necessary to add heat to the water or to do work on it (say, by stirring). The amount of heat or work needed to raise the temperature of 1 gram of water by 1° Kelvin (or Celsius) is called the "specific heat of water" or the "heat capacity of water," and it changes slightly depending on the initial temperature of the water.

The entropy of 1 gram of water can be experimentally measured in the following way. Start with the water at a convenient initial temperature and pressure. Measure the heat needed to raise its temperature by 1° Kelvin; this is the specific heat (at constant pressure). Divide the specific heat by the initial temperature, and call this number x_1. Next measure the heat needed to raise the water temperature by one more degree Kelvin, divide that by the temperature, and call that number x_2. Continue until the temperature of interest is reached. The change in entropy of the water between the initial and final states is approximately equal to the sum of the xs, times 1° Kelvin to adjust the units.[7]

In this way, the entropy change of different materials can be experimentally determined. As mentioned above, "absolute," or "standard," entropies are defined with an initial condition of 0°K and one atmosphere pressure and a final condition of 300°K and one atmosphere pressure. For example, the standard entropy of one mole of copper (63 grams) is 33 J/°K = 8 cal/°K, where 'cal' denotes calories (Rao, 1985: table C–1).

To demonstrate how chemical engineers use entropy, consider a reaction $A + B \rightarrow 2C$ where A, B, and C are ideal gases. Most chemical reactions do not fully go "to completion," which is to say that usually some of the reactants do not combine to form products but instead remain in their initial form. Suppose one started with 1 mole of A and 1 mole of B and suppose one knew from tables the standard entropies of A, B, and C. Standard entropies are denoted S^o. Then when 1 mole of A and 1 mole of B combine to form C, the entropy change of the system would be $2S_C^o - S_A^o - S_B^o$. Suppose one also knew from tables the "standard heat of formation" H^o for A, B, and C, which is the amount of heat involved when 1 mole of A, B, or C is formed from its respective constituent elements.[8] Then when 1 mole of A and 1 mole of B combine to form C, $2H_C^o - H_A^o - H_B^o$ is the heat ejected to the surroundings, and equation (5.1) gives the surrounding's entropy change.

Suppose n_A is the number of moles of A which actually do combine with B to form C. Gaskell (1981) writes:

> Starting with 1 mole of A and 1 mole of B (i.e., 2 moles of chemicals), as 1 atom of A reacts with 1 atom of B to produce 2 molecules of C, then at any time, [the number of moles of A,] n_A [is equal to the number of moles of B,] n_B, and [the number of

moles of C is given by] $n_C = 2 - n_A - n_B$ [which is equal to $2 - n_A - n_A$, which in turn] $= 2(1 - n_A)$.

Using these relationships, from (5.3) and (5.1) the total entropy change due to $A + B \rightarrow C$ is:

$$\Delta S_{total} = \Delta S_{configurational} + \Delta S_{thermal,\ system} + \Delta S_{thermal,\ surroundings}$$

$$= -R\left[n_A \ln \frac{n_A}{n} + n_B \ln \frac{n_B}{n} + n_C \ln \frac{n_C}{n} \right]$$

$$+ \left[\left(n_A S_A^\circ + n_B S_B^\circ + n_C S_C^\circ \right) - \left(1S_A^\circ + 1S_B^\circ \right) \right]$$

$$+ \frac{1}{T}\left[\left(n_A H_A^\circ + n_B H_B^\circ + n_C H_C^\circ \right) - \left(1H_A^\circ + 1H_B^\circ \right) \right]$$

$$= -R\left[n_A \ln \frac{n_A}{2} + 2(1 - n_A)\ln(1 - n_A) \right]$$

$$- (n_A - 1)\left(2S_C^\circ - S_A^\circ - S_B^\circ \right)$$

$$- \frac{1}{T}\left[-(n_A - 1)\left(2H_C^\circ - H_A^\circ - H_B^\circ \right) \right]. \tag{5.6}$$

Equilibrium occurs in the state of maximum entropy, since from there, any deviation would decrease entropy and thus not be allowed by the Entropy Law. The state of maximum entropy is found by maximizing ΔS_{total} with respect to n_A. Call the maximum point n_A^*. Since we started with exactly 1 mole of A, and since n_A^* moles of A combine with B to form C, $1 - n_A^*$ gives the percent completion of $A + B) \rightarrow C$. In this way, the Entropy Law can be used to calculate the equilibrium percent completion of any chemical reaction.

A common process in extractive metallurgy is the separation of two chemically combined elements, the pure form of one of which is economically valuable. As an introduction to such reactions, consider breaking the chemical bonds of mercury oxide, briefly considered in note 2:

$$HgO \rightarrow Hg + \tfrac{1}{2}O_2. \tag{5.7}$$

At 25°C, for each mole of HgO, this reaction causes the entropy of the system to change by + 108 J/°K and the entropy of the surroundings to change by –305 J/°K, so the total entropy change is – 19 7 J/°K and (5.7) will not occur. On the other hand, at 300°C, this reaction causes the entropy of the system to change by +207 J/°K and the entropy of the surroundings to change by –197 J/°K, so the total entropy change is +10 J/°K and (5.7) will occur. (The data is from Keenan *et al.*, 1976: 427.) Put another way, at 25°C, $Hg + \frac{1}{2}O_2 \rightarrow$ HgO increases entropy, while at 300°C, $HgO \rightarrow Hg + \frac{1}{2}O_2$ increases entropy. Therefore chemical combination does not always increase entropy, and chemical separation does not always decrease it. Above (p. 92) we found that physical mixing does not always increase entropy, and physical separation does not always decrease it.

Mercury is found in nature as HgS, mercury sulfide; when roasted, the reaction $HgS + O_2 \rightarrow Hg + SO_2$ increases entropy, and so can be used to recover mercury. At room temperature, $Hg + \frac{1}{2}O_2 \rightarrow HgO$ increases entropy, so one may wonder why mercury does not immediately "rust" when it is cooled down after the $HgS + O_2 \rightarrow Hg + SO_2$ reaction (ibid.: 427, 497). The reason is that pure Hg at room temperature is "metastable" in the sense defined above (p. 89). The upshot is that if a process would result in a decrease in entropy then it can never occur, but the converse statement is false. (In other words, if a process never occurs, one cannot conclude that that process would result in a decrease in entropy; there might just be an activation energy barrier standing in the way of the stable equilibrium.) As another example of entropy calculations, consider:

$$Cu_2S + O_2 \rightarrow 2Cu + SO_2 + energy, \qquad (5.8)$$

which describes how copper is commercially purified in a Peirce–Smith converter at approximately 1500°K and 1 atmosphere pressure. At this temperature and pressure, the entropies of the chemicals involved in the reaction are the following, in cal/°K: 66.27 for 1 mole of Cu_2S, 61.653 for 1 mole of O_2, two times 20.812 for the 2 moles of Cu, and 78.470 for 1 mole of SO_2. The entropy of the system hence changes by -7.829cal/°K if (5.8) goes to completion from left to right; this is roughly equal to -33J/°K. The entropy change of the system's surroundings is found by calculating the heat generated by the reaction, then using (5.1). The heat generated by the reaction is found by subtracting the enthalpies of the reactants (41 338 J for Cu_2S and 40 597 J for O_2) from the enthalpies of the products (94 609 J for 2Cu and -234 406 J for SO_2). Doing this yields -221 732 J, so the heat lost by the system, and gained by the environment, is 221 732 J. (This is the "energy" term in (5.8).) According to (5.1), the environment's heat gain causes the environment's entropy to increase by the heat absorbed divided by the temperature (1500°K), which yields 148J/°K. This increase in the entropy of the environment more than offsets the 33J/°K decrease in the entropy of the system. Reaction (5.8) owes its spontaneity to this fact that it increases entropy; in turn, this spontaneity is what makes the commercial purification of copper possible.[9]

It is hardly possible to exaggerate the importance of maximum-entropy calculations like these to the chemical and metallurgical industries: they are the only means of predicting the direction in which chemical reactions will proceed.[10]

5.6   ENTROPY AND ENERGY

There is no such thing as "low-entropy energy." Configurational entropy clearly applies only to matter. Thermal entropy, from equation (5.1), does involve heat (which is a flow of energy), but the thermal entropy is that of the material which is absorbing or emitting heat. Only matter, not energy, has entropy.

This is not to say that all energy is stored in ways that are equally useful. Gravitational and electromagnetic potential energy, and organized kinetic energy such as in a spinning flywheel, are forms of energy called "ordered energy" that can be completely converted into work, whereas chemical energy, thermal radiation (photons), and the thermal energy of matter (heat) are "disordered energy" which cannot be completely converted into work (Kotas, 1985: 30).

Suppose a hot body at temperature T_2 is placed inside a house of temperature T_1 while the outside air is at $T_0 < T_1 < T_2$. If some of the body's heat energy is allowed to flow to the room air, some ability to do work is lost. If heat energy Q had not gone to the room air, the maximum amount of work it could have done would have been $Q(1 - (T_0/T_2))$ as we saw in section 5.2. Once Q has gone to the room air, the maximum amount of work it can do is only $Q(1 - (T_0/T_1))$. The "maximum amount of work possible" has fallen by the difference between these two quantities, which after some manipulation is:

$$T_0\left(\frac{Q}{T_1} - \frac{Q}{T_2}\right) = T_0\Delta S_{universe}$$

(Zemansky, 1968: 237). It turns out that this is a general result for any irreversible process. The result is traditionally expressed in the following way: if T_0 is the temperature of the coldest very massive body (heat reservoir) which could be put into contact with a system, then

> the *energy* that becomes *unavailable* for work during an irreversible process is T_0 times the *entropy change* of the universe that is brought about by the irreversible process ... Since irreversible processes are continually going on in nature, energy is continually becoming unavailable for work. This conclusion, known as the *principle of degradation of energy* and first developed by Kelvin, provides an important physical interpretation of the entropy change of the universe. (Ibid.: 238; emphases changed)

Using the traditional symbols, $E = T_0 \Delta S$, or more precisely:

$$\text{energy becoming unavailable for work} = T_0 \Delta S_{universe} \qquad (5.9)$$

(ibid.: 238–9). An examination of Zemansky's proof shows that (5.9)'s left-hand side refers only to *disordered* energy (so perhaps a better expression of the relationship would have been "the loss of 'Q which could have been turned completely into work' is equal to $T_0 \Delta S$"). There is no law of degradation of kinetic energy, for example. Kelvin's principle is said to foreordain the eventual "*heat* death" of the universe (Keenan *et al.*, 1976: 420).

It should be clear that (5.9) does not contradict our assertion that entropy is a property of matter, not a property of energy.[11]

The way energy and entropy interact in photosynthesis is instructive. Solar energy falling on the Earth is capable of doing work; it is non-degraded. In an irreversible process, this energy is captured by a plant, then it is ejected by the plant to the plant's surroundings. This ejection of energy degrades the energy. Call the energy degraded E_1. Then from (5.9):

$$E_1 = T_0(\Delta S_{plant} + \Delta S_{plant's\ surroundings}). \qquad (5.10)$$

This equation allows the plant's entropy to decrease as long as the plant's surroundings' entropy increases more, so (5.10)'s right-hand side remains positive – and this is precisely what happens. The degrading flow of energy allows the maintenance of a low-entropy structure (the plant).[12] It is only a small step from this to claim that the flow of energy allows the *creation* as well as the maintenance of low-entropy structures. If we change the word "allows" to "requires," we only slightly overstate the world view of post-Prigogine writers such as Wicken (1987) or Fry (1995), for whom degrading flows of energy become wonderfully creative: "neither is it possible for a dissipative process to occur without *producing* some transient nonequilibrium *structure*" (Wicken, 1987: 115, emphasis added; see also fig. 10-2).

Before leaving this chapter, mention should be made of one important physical result which is not part of thermodynamics: the conservation of matter. As Georgesçu-Roegen wrote (*Energy and Economic Myths*, 1976: 11–12):

despite the Einstein equivalence of mass and energy, there is no reason to believe that we can convert energy into matter except at the atomic scale in a laboratory and only for some special elements. We cannot produce a copper sheet, for example, from energy alone. All the copper in that sheet must exist as copper (in pure form or in some chemical compound) beforehand.

NOTES

1. A classic reference for heat engines is Zemansky (1968: §9–4 and p. 235). The best simple explanation of the maximum-efficiency equation appears in Lodge (1929: 39–40): "If the quantity of heat energy Q is lost by a body at temperature T, and gained by a body at temperature T', then the entropy lost is Q/T, and the [entropy] gained is Q/V'. The gain is

greater than the loss, if T' is less than T. But if some of the heat is utilised, – converted into mechanical or other forms of energy by means of some kind of engine, – so that only Q' is imparted to the colder body (the work done corresponding to the difference Q – Q') then it may be that Q/T = Q'/T'; and in that case the entropy of the system remains constant. This is the condition for a perfect engine conducting a reversible operation. Irreversible operations increase entropy or dissipate energy, for the lost availability cannot be recovered. The efficiency of an engine means the ratio between the portion of heat utilised and the heat taken from the source. It is (Q – Q')Q; and we see now that in a perfect engine this would be T – T')/T, because the Q's and the T's are then proportional."

2. There are cases when chemical "mixing" does not even increase *configurational* entropy. Consider the reaction of mercury and oxygen to form mercury oxide at 500°C. This "mixing" reaction is written $Hg(gas) + \frac{1}{2}O_2(gas) \rightarrow HgO(solid)$. Configurational entropy decreases appreciably when the 2 moles of gas turn into 1 mole of solid. In fact, though this reaction is exothermic (it would make the environment's entropy rise), the decrease in configurational entropy is so great that, taken as a whole, the reaction decreases entropy. Hence HgO spontaneously decomposes at that temperature. See Keenan *et al.* (1976: §17–8.4).

3. See Cottrell (1967: 77–9) and Miller (1976: 148–55, 158f (23d)).

4. As pointed out above, the "unattainability'" form of the Third Law of Thermodynamics is that "it is impossible to reach absolute zero by any finite number of processes" (Dugdale, 1996: 177). However, laboratory refrigerators can get down to 0.03°K (ibid.: 171), and experimentalists then use extrapolation to infer behavior at absolute zero. Using specialized techniques, the lowest temperature obtained as of 1968 was 0.0000012°K (Zemansky, 1968: 485).

5. The reason that we put the word probability in quotes is that the precise relationship between probability and statistical mechanics is somewhat problematic; see Lozada (1995).

6. A paraphrase of Atkins' (1984: 157–60) discussion of oil and water is: When a hydrocarbon is surrounded by water, the water molecules form a delicate molecular cage around it. The formation of the cages corresponds to a significant decrease in configurational entropy – so significant that it overwhelms all other contributions to the total entropy change. Hence the spontaneous direction of change is from dispersed oil molecules to a blob of oil.

7. The exact formula is $S = \int_i^f (C_p(T)/T)\, dT$ where i denotes some convenient initial reference state, f denotes the state of interest, and C_p denotes the "heat capacity at constant pressure."

8. The technical term for H is "enthalpy." According to Zemansky (1968: 276), "the change in enthalpy during an isobaric process is equal to the heat that is transferred." Zemansky adds that isobaric processes are very important in engineering and chemistry. We assume that the chemical reaction we are studying takes place isobarically.

9. The physical data for Cu come from Pankratz (1982: 129), and for O_2 from ibid. (277). For SO_2, (ibid.: 348) lists S^o and $H^o - H^o_{298}$, but H^o_{298} is supplied from table C-1 of Rao (1985). For Cu_2S, Pankratz *et al.* (1987: 102) lists S^o and $H^o - H^o_{298}$, but H^o_{298} had to be supplied from table C-1 of Rao (1985). There are 4.184 Joules in 1 calorie and 4184 Joules in 1 kilocalörie.

    It turns out that $Cu_2S + O_2 \rightarrow 2Cu + SO_2$ increases entropy even at room temperature. The quickest verification is in the ΔG^o column of King *et al.* (1973: table 70). (For the interpretation of ΔG^o see Lozada, 1999.) Why, then, are most copper deposits remaining in the US in the form Of Cu_2S instead of Cu? The reason is that Cu_2S is metastable, albeit not stable, at room temperature.

10. Texts in chemistry and metallurgical thermodynamics usually present maximum-entropy calculations using "Gibbs Free Energy" $G = H - TS$ instead of entropy, but the results are the same. See the Appendix to Lozada (1999). That paper also explains "exergy," which is simply the sum of the first-law energy losses and the second-law energy degradations described in (5.9) of section 5.6.

11. It is possible to calculate the entropy of a collection of photons, which have zero mass (Zemansky, 1968: §13–13). However, the entropy is still considered a property of the photons as particles.

12. We explain why life implies lower entropy than death in Chapter 6's note 1.

6. Thermodynamics and Georgesçu-Roegen's economics

6.1 INTRODUCTION

The second section of this chapter is a straightforward application of Chapter 5 to problems of economics. Sections 6.3–6.5 pick up controversial topics related to this area. Section 6.3 treats Georgesçu-Roegen's proposed "Fourth Law" of Thermodynamics, while sections 6.4 and 6.5 treat Boltzmann's H-theorem, the relation between energy and entropy, and other topics which can cause confusion. The last section covers epistemology, methodology, and Georgesçu-Roegen's opinions of the state of neoclassical (or, as he preferred to call it, "standard") economics.

6.2 ENTROPY AND THE ECONOMY

The first implication of the Entropy Law for economics is its establishment of a maximum efficiency for heat engines (for example, internal combustion engines and generators of electric power). We described this in section 5.2.

To derive the second implication, suppose humans use a material X as a source of energy. For example, X could be uranium or coal. In the process of obtaining energy from X, entropy increases and energy is degraded. If people were to try to recycle the products of the process, thereby reconstituting X, the First Law implies that at least as much energy would be locked up in the recycling as was obtained when X was first used. The Second Law demands that the entropy of the universe increase during the recycling. That means that during recycling the sum of the energy that has to be degraded and the energy that has to be locked up is strictly greater than the amount of energy which was obtained from X in the first place, and it is strictly greater than the amount of energy which will be obtained from X once it has been reconstituted. So while the First Law of Thermodynamics says that energy can be neither created nor destroyed, the Second Law tells us that recycling energy sources will always be uneconomical.

A third implication of the Entropy Law is that if a collection of non-interacting particles has to be sorted – for example, if Cu_2S has to be separated from waste rock – then thermal entropy is not involved and the configurational entropy of the system has to decrease by (5.3) or its appropriate generalization. (See, for example, Faber *et al.*, (1995: ch. 4.)) Therefore an amount of energy given by (5.9) would have to be degraded regardless of future technological change. Identical considerations apply to recycling many materials. After all, scarcity of an element like copper, whose terrestrial stock is essentially constant, can only mean that copper becomes dispersed as it is used. "Scarcity" thus refers to the difficulty of decreasing configurational entropy – difficulty which will always include a minimum energy requirement, because of the Second Law. (Here the term "energy" means energy in the physical sense, not to be identified solely with materials which are called energy commodities in the economic sense. See section 6.4 below.)

Moving to the level of economic theory, Georgeşçu-Roegen (in *The Entropy Law and the Economic Process*, 1971: 18, 282) gives the example of poisonous mushrooms to show that having "low" entropy is not a sufficient condition for having economic value. This shows that an entropy theory of value must be incorrect. Georgeşçu-Roegen did however contend that "low" entropy is a necessary condition for economic value.

It is important not to read too much into what Georgeşçu-Roegen contended. He did not write that more economic value implies lower entropy. One reason why such a contention would be false is that, because entropy is an extensive concept, 1 kilogram of copper has lower entropy than 2 kilograms of copper – yet 1 kilogram of copper cannot be more valuable than 2 kilograms of copper. One might try to solve this by using an intensive measurement of entropy: joules per degree Kelvin *per kilogram*, or perhaps *per mole*, or (less likely) *per liter*. However, the contention would still be wrong. Adding impurities to iron to create steel simultaneously increases the entropy of the iron and increases its economic value. Similarly, entropy and economic value both increase when a hunter kills a flying duck in order to sell it in a market.[1] There are other cases in which economic value increases while entropy does not change. An example would be the stamping of thin copper cylinders to make them into pennies: a calorimetric measurement of entropy would not show any difference between the stamped and unstamped copper. Another example would be in the provision of services (such as life insurance or home mortgages) which have economic value but do not involve physical commodities and hence do not involve entropy. Finally, there are of course production processes which decrease entropy and increase economic value, such as the purification of many metals.[2]

With these examples in mind, precisely what does it mean to say that "low" entropy is a necessary condition for economic value? We could try asserting that "all inputs to the economic system must have low entropy." This is attractive because even in an example such as the manufacture of steel, decreases in entropy had to occur (to purify the iron and the chemicals which would later be added to the iron) in order to make the final entropy-increasing step possible. However, it is not easy to give an appropriate arithmomorphic definition of the phrase "low entropy."

Leaving aside the question of commodity value, we can at a minimum say that as a historical fact (perhaps not as a logical necessity, but who knows?), industrial civilization requires the transformation of large quantities of materials from their naturally occurring states. This transformation requires either decreases in entropy (if the naturally occurring state is a thermodynamically stable equilibrium) or increases in entropy which without an appropriate catalyst or change in physical environment would occur extremely slowly or not at all (if the naturally occurring state is a thermodynamically metastable equilibrium). Otherwise nature would have already accomplished the transformation. The ore Cu_2S is thermodynamically metastable, so recovering copper from it is an example of the second type of process. The ore bauxite contains Al_2O_3 (alumina), which is thermodynamically stable, so recovering aluminum from it is an example of the first type of process. (Once aluminum is purified, its entropy spontaneously increases again as a thin film of Al_2O_3 forms on its surface almost immediately.)

The core insight is that all economic transformations, regardless of their type, must result in the entropy of the universe rising – otherwise the transformation could not occur. The passing of each moment leaves us in a universe of ever-greater entropy.

Georgesçu-Roegen tried to capture this idea by saying that "low entropy continuously and irrevocable dwindles away" and "our accessible environment is like an hourglass . . . in which the useful matter-energy from the upper half turns irrevocably into waste as it continuously pours down into the lower half" (*Energy and Economic Myths*, 1976: 15, xvi). In making this hourglass analogy he was influenced by Schrödinger's comment that living things sustain themselves on a flow of low entropy (see *The Entropy Law and the Economic Process*, 1971: 10, 193). The objections given a few paragraphs ago are answered by positing that the set of continually degrading matter is not identical with the set of matter passing through the economic system.

There is still, however, a severe conceptual difficulty with the "low entropy as sand in an hourglass" analogy – a difficulty which, perhaps ironically, we can spot only because of Georgesçu-Roegen's stubborn and brilliant insistence on the distinction between cardinal and weakly cardinal magnitudes (as defined in section 5.2). The analogy depends on being able to make statements such as, "let S denote the present stock of terrestrial low entropy" (*Energy and Economic Myths*, 1976: 58). Yet only *changes* in entropy levels are defined, not absolute levels of entropy. With no absolute level of entropy, who is to say that the constant degradation of matter, from low to high entropy, poses a constraint? If we can always redefine the amount of sand in the top of the hourglass to be whatever level we please, will the hourglass not run forever?

While not explicitly posing this question, Georgesçu-Roegen ("Matter matters, too," 1977: 299–300) implicitly proposed a solution to it, based on the Helmholtz free energy $F = U - TS$ (not to be confused with the Gibbs free energy mentioned in note 10, Chapter 5). We explore his idea, which we find ultimately unconvincing, in a note.[3] Another potential solution which is ultimately unavailing is to appeal to Kelvin's insight that the flow of entropy from low to high entails a simultaneous degradation of energy (equation (5.9)). The difficulty is that it is not clear how to measure the stock of "non-degraded energy."[4]

A better analogy than Georgesçu-Roegen's hourglass can be developed starting from the fact that it is possible to describe the thermodynamically stable equilibrium state of any isolated system. For example, we found the thermo-dynamically stable equilibrium state of the hypothetical system $A + B \Leftrightarrow C$ in section 5.5. As another example, in the stable equilibrium state of the solar system, the solar system will have no temperature differences, and all entropy-increasing chemical reactions will have occurred. Given this description of the thermodynamically stable equilibrium state of a system, we can compute the entropy difference between that state and the present state. This entropy difference is the maximum entropy change which can occur in the system (because once the stable state is reached, any change would by definition decrease the system's entropy). This maximum entropy change is what Clausius had in mind when he said that "the entropy of the universe tends to a maximum" ("Thermodynamics and we, the humans", 1993: 188).

An appropriate analogy, then, is of an hourglass with three instead of two chambers. The bottom and top chambers begin by being full of sand, while the middle chamber is empty. The size of the middle chamber represents the maximum ΔS of the system, which has a definite size. The sizes of the bottom and top chambers are arbitrary, since the stock of entropy is defined only up to an arbitrary additive constant. (The top chamber has to be at least as big as the middle chamber, or else it could not undergo the ΔS which it is assumed to be able to undergo.) As time goes on, sand can drop from the top chamber to the middle one, but once the middle chamber becomes full, no more sand can drop and no more changes can take place. This represents the point at which the entropy change has reached its maximum, and no further entropy increase is possible. What stops the process is not a lack of sand, but a limitation on how many grains of sand can move.

This three-chambered hourglass analogy is more faithful to the underlying physics than Georgesçu-Roegen's two-chambered hourglass. The implications for economics, however, are not very different. Industrial civilization requires many transformations of materials. Each transformation increases entropy, and hence moves the solar system faster to its state of maximum entropy – a state which will prevent any further decrease in entropy, and hence any further transformations of materials, from occurring. Georgesçu-Roegen also felt that the ever-closer we approach the state of maximum entropy, the ever-harder it will be to perform our desired transformations, which is a plausible and interesting conjecture.

Georgesçu-Roegen's vision is not an arithmomorphic statement about the mechanics of price determination or value theory. It is a sweeping dialectical contention about the past, present, and future requirements of industrial civilization, and the prospects for such civilizations decades and centuries hence (*The Entropy Law and the Economic Process*, 1971: 19).

6.3   THE FOURTH LAW

Barring exchanges between matter and energy according to $E = mc^2$, the amount of matter in the universe is constant. The amount of matter on Earth is very close to constant (the exceptions are meteorites and items sent into space). The amount of each element on Earth is also very close to constant (the exceptions are nuclear reactions). In this section we address whether there are any other fundamental physical constraints to the use of materials.

Zemansky (1968: 193) writes:

> It is a matter of everyday experience that dissipative effects, particularly friction, are always present in moving devices. Friction, of course, may be reduced considerably by suitable lubrication, but experience has shown that it can never be completely eliminated. If it could, a movable device could be kept in continual operation without violating either of the two laws of thermodynamics. Such a continual motion is known as *perpetual motion of the third kind*.

Georgesçu-Roegen ("Matter matters, too," 1977: 303; "Energy, matter, and economic valuation", 1981: fn. 16) defines perpetual motion of the third kind in the same way: "a closed system that can perform *forever* work between its subsystems." ("Closed" means that matter cannot enter the system but energy can.)

Georgesçu-Roegen's Fourth Law of Thermodynamics is that perpetual motion of the third kind is impossible (ibid.). Alternatively, it holds that complete recycling is impossible ("Energy, matter, and economic valuation", 1981: 60).

Mayumi (1993) thinks the Fourth Law is incorrect. So do many biologists, who are familiar with ecological systems in which recycling of myriad materials seems to be complete. The biologists disagree with Georgesçu-Roegen (*The*

Entropy Law and the Economic Process, 1971: 302) when he contends that buffalo grazing in the wilderness with no human influence will degrade their environment. They see the biological and geological history of Earth as being made up of long periods in which change was imperceptible and hence recycling must have been, at least for all practical purposes, complete. This did not convince Georgesçu-Roegen, who had even longer time periods in mind.[5] He believed that in these very long time periods, even "Forests cannot be turned into everlasting suppliers of appreciable amount of wood" ("Looking back", 1991: 15).

The biologists' claim of perfect recyclability by nature has not been refuted experimentally. It may be. The question remains open. So does the question of whether, say, an electromagnet could release iron pieces and then pick them all up again infinitely many times (see Georgesçu-Roegen, ("Thermodynamics and we, the humans", 1991: 198). Even if the Fourth Law is incorrect, however, with current technology, material recycling is far from complete, and so valuable materials are constantly being dissipated into forms which cannot be reused.

6.4   CONTROVERSIES CONCERNING ENERGY AND ENTROPY

Georgesçu-Roegen wrote that the Second Law of Thermodynamics "has been, and still is, surrounded by numerous controversies – which is not at all surprising" since "this concept is so involved that one specialist judged that 'it is not easily understood even by physicists'" (*The Entropy Law and the Economic Process*, 1971: 128, 147 fn. 10; *Energy and Economic Myths*, 1976: 7). Here we investigate some of the difficult issues.

Georgesçu-Roegen's treatment of the relationship between energy and entropy went through three stages in the 1970s:

1. Entropy is energy dissipation and has nothing to do with matter. (We know that, actually, matter has entropy, and an increase in the entropy of matter is accompanied by energy dissipation.)
2. The Entropy Law implies that matter dissipates. (We know that, actually, the Entropy Law implies that energy dissipates: configurational entropy may go up but it need not, because what counts is configurational entropy plus thermal entropy.)
3. There are two entropies, one for energy and one for matter; the Second Law deals with the former, and the Fourth Law deals with the latter. (We know

that, actually, there is only one entropy in physics – but the Fourth Law still might be true.)

Stage 1 is represented by the first part of Georgesçu-Roegen's 1971 book *The Entropy Law and the Economic Process*. He first defined entropy in terms of energy degradation (p. 129), referring to 1 strictly in passing. As far as p. 193 of this book, Georgesçu-Roegen did not acknowledge that entropy is only a property of materials.[6]

Stage 2 starts in the middle of Georgesçu-Roegen's 1971 book and continues through 1976. On p. 277 of the 1971 book, he wrote: "*our whole economic life feeds on low entropy*, to wit, cloth, lumber, china, copper, etc., all of which are highly ordered structures" with no intervening explanation of how entropy could switch from being a property of energy to being a property of materials, or of order. Indeed, earlier in the book he very strongly objected to the ideas of Boltzmann's statistical mechanics, with which "entropy became associated with *the degree of disorder* (however defined): (p. 159). Later in this period, Georgesçu-Roegen wrote quite explicitly that "the Entropy Law in its present form states that *matter, too, is subject to an irrevocable dissipation*" (*Energy and Economic Myths*, 1976: 8).[7]

Stage 3 began in 1977, when he realized the difficulty with stage 2's position ("Matter matters, too," 1977: 301–2):

On what *operational* basis can the loss of matter availability be treated as being of the same essence as the loss of energy availability? In other words, why should the sum of the two entropies

Entropy of Energy Diffusion + Entropy of Matter Mixing

have one and the same meaning regardless of its distribution among the two terms?

In classical thermodynamics, the answer to Georgesçu-Roegen's question is that there is only one entropy, defined by (1). Among (1)'s implications are (3), which deals with matter mixing, and (9), which is Kelvin's "principle of degradation of energy". In statistical thermodynamics, the answer to Georgesçu-Roegen's question is that "thermal entropy" and "configurational entropy" have the same meaning because this assumption yields predictions which accord with experiments. Because Georgesçu-Roegen did not understand the classical answer to his question[8] and rejected the statistical mechanical answer (as he rejected all of statistical mechanics), the only way he could take both manifestations of entropy into account was to postulate that there were two entropy laws, the Second and the Fourth, one for each manifestation of entropy.[9]

We can unequivocally state that Georgesçu-Roegen's Fourth Law is the wrong answer to the question it was partially formulated to answer: the relationship between energy and entropy (because as mentioned above that question already has satisfactory answers in both classical and statistical thermodynamics). Clearly, Georgesçu-Roegen made some statements which in retrospect were unfortunate.[10] He was not infallible. Neither, as we shall see below, were Boltzmann, Planck, or Shannon.

What must not be forgotten is that, as we have argued earlier in this chapter, despite Georgesçu-Roegen's errors, his conclusions about the import of the Entropy Law for economics were broadly correct. Even the Fourth Law might still be true, despite being the wrong answer to the question it was partially formulated to answer. Georgesçu-Roegen referred in another context to "the truth that correct conclusions may very well be based on wrong premises" ("Matter matters, too", 1977: 311 fn. 6). His rare mistakes sometimes, though of course not always, illustrate that truth.

Georgesçu-Roegen is not the only important economic analyst whose treatment of energy and entropy contained problematic elements. Here is the treatment by Chapman and Roberts (1983: 10, 24–5):

> To an engineer or scientist there is something which, if available in unlimited quantities, could, at least in principle, resolve all material resource issues. This is energy, or more properly, highly available energy of the sort released in the conversion of fossil and fissile fuels . . . All industrial production processes involve the degradation of the highly available energy, obtained from fuels, to low grade heat energy. Each and every stage of production can be described in these terms . . . There are two features of this physical mode of description that are significant for our discussions: these are that there is a definite theoretical limit to the minimum energy input to a process and that there is no substitute for this minimum energy input . . .
>
> In the summer of 1976, an international group of economists and science-based energy analysts met in Sweden to discuss the relationship of economic and physical approaches to resource issues. Early in the meeting, it became clear that there was a deep division between the two groups. To the economists, it seemed strange to focus so much attention on one of the inputs of production . . . [Tjalling] Koopmans proposed a test which focused on the essential issue, "What makes energy special?" . . .
>
> The solution to this test, put forward by Chapman at the meeting, lay . . . in the non-substitutability and minimum input requirements of energy. It is possible to conceive of a process for producing iron from iron ore without any lime, or carbon, or water, or even labor – but it is impossible to affect the transformation without an input of energy. At the meeting, this exchange altered the whole approach to the issues. The economists were genuinely surprised and fascinated to know that there were thermodynamic statements that could be made that were process independent.

This "surprising" realization by economists came to be reflected in, for example, Dasgupta and Heal's classic text (1979: 208, §7.3).

One problematic aspect of this passage is that, if taken at face value, it implies that it is possible to conceive of a process for producing iron without iron ore, just using energy. We disagree, as discussed in the last paragraph of Chapter 5.

The second problematic element is that while it may not be possible to produce iron without an input of energy, it is quite possible to at least partially purify some other metals without an input of energy – or at least without energy as an economist would recognize it.

One example of this is the purification of materials which in nature are thermodynamically metastable, such as Cu_2S. As we saw in section 5.5, it is quite possible to produce copper from Cu_2S without an input of energy. Most of the copper supplied to the world today is produced in just this way, using (5.8). That chemical reaction not only creates an energy output, it actually creates an energy output large enough to keep the temperature of the materials at the appropriate, very high level, and also enough to generate electricity as a by-product. The only flow inputs to that reaction are the copper ore and oxygen in the air (though sometimes companies add extra oxygen).

Another example in which a metal is partially purified without a fuel input is the production of partially purified copper from oxide copper ores in Africa. Even though CuO is thermodynamically stable, not merely metastable, at room temperatures, partially purified copper is produced from it using only sulfuric acid (Biswas and Davenport, 1980: 254–5, 258). A typical reaction is $CuO + H_2SO_4 \rightarrow Cu^{++} + SO_4^- + H_2O$ (Biswas and Davenport, 1994: 15).

The difficulty in the Chapman–Roberts approach which these examples reveal is the potential for confusing energy with fuels. Because of the Second Law, all physical transformations increase entropy if the system boundaries are drawn large enough for the system to be isolated. This entropy increase in turn implies a degradation of energy according to Kelvin's principle mentioned earlier (5.9). However, we have seen that the necessary energy can be supplied by chemical changes in non-fuel materials. In the processing of copper sulfide, the necessary energy is supplied by what is best described as simply burning the copper sulfide. In the leaching of oxide copper ores, the necessary energy is supplied by the interaction of sulfuric acid with the ore. In both these copper-purifying processes, useful energy in the abstract scientific sense, which includes energy implicit in chemical bonds, must have decreased because of the Second Law, but energy in the form of a tradable economic commodity has been produced in the first case, and has not been consumed in the second. Of course, there are other purification processes, such as the production of aluminum, which are net consumers of "energy in the form of a tradable economic commodity." Nevertheless the conclusion is that minimum required energy degradation is not the same as a minimum required fuel input. There is no such thing as a minimum fuel requirement which applies to all industrial processes because of the Second Law. Georgesçu-Roegen ("Matter matters, too," 1977) could not have been more correct in saying that "Matter matters, too."[11]

6.5   OTHER CONTROVERSIES CONCERNING ENTROPY

One controversy concerning entropy is very basic: What is the definition of entropy in statistical mechanics? In Boltzmann's original work, instead of the correct (5.4), the relation between S and W is given by $S = k \ln W$. This is also asserted by many others, including Planck (1923) and the standard text of Zemansky (1968: 260). The relation $S = k \ln W$ implies that absolute entropy is defined in statistical mechanics even though it is not defined in classical thermodynamics. Zemansky teaches precisely this in some places (ibid.: 266, §10–5) but, quoting Fowler and Guggenheim, Zemansky teaches the opposite in other places (ibid.: 241, §9–12).

As we know from persuasive arguments such as those of Dugdale (1996), the correct formula is (5.4), and statistical mechanics does not define absolute entropy.[12] Georgesçu-Roegen (*The Entropy Law and the Economic Process*, 1971: 146) was on the correct side of this argument, while Max Planck was not. Georgesçu-Roegen correctly criticized Boltzmann's original result on the grounds that it implied that entropy was a cardinal variable, whereas entropy is actually only a weakly cardinal variable.

Making this correction, from a modern point of view there is one and *only* one result in Boltzmann's statistical mechanics which is wrong. This result is known as the H-theorem. Georgesçu-Roegen was on the right side of this debate also, as *The Entropy Law and the Economic Process* contains many harsh criticisms of the H-theorem.

To summarize the issues briefly, Boltzmann's H-theorem is the claim that time-reversible Newtonian mechanics can generate the arrow of time and, in particular, that it can generate the time-irreversible Second Law of Thermodynamics, namely the increasing entropy of an isolated system. Boltzmann's "proof" of his theorem was strongly questioned in the nineteenth century, and this criticism may have played a role in Boltzmann's suicide. For most of the twentieth century, including the 1960s and early 1970s when Georgesçu-Roegen was writing *The Entropy Law and the Economic Process*, the H-theorem was accepted as true. Now, however, the consensus among philosophers of science is that the H-theorem is false, for many of the same reasons that Georgesçu-Roegen brought up (though there is no evidence that any of the philosophers read Georgesçu-Roegen.)[13]

The reason for the vehemence, even fury, that one can see in Georgesçu-Roegen's criticisms of the H-theorem is epistemological. He felt mechanical models are like the Sirens, seductively luring economists to their doom. Mechanical models, such as the Newtonian mechanics which the H-theorem says is the basis for thermodynamics, essentially imply that the Second Law's arrow of time, and with it the notion of qualitative change, of the emergence of genuine novelty, and of the irreducible unpredictability of history, are all illusions. The H-theorem would have us believe that the world is, at its most basic level, purely mechanical. So the issue is "extremely important, for it pertains to whether the phenomenal domain where our knowledge is both the richest and the most incontrovertible supports or denies the existence of evolutionary laws" (*The Entropy Law and the Economic Process*, 1971: 169). The falsity of the H-theorem shows us that there is another epistemological path besides the mechanistic one. That alternative path is the one Georgesçu-Roegen believes we have to take in order to understand fully the evolution of our economy, as we discuss in section 6.6 below.

Together with Georgesçu-Roegen's wholly justified rejection of Boltzmann's H-theorem, he decided to reject the rest of Boltzmann's statistical mechanics as well (see, for example, *The Entropy Law and the Economic Process*, 1971: 7, 168). This was unwise. As discussed in section 5.4, statistical mechanics has provided accurate predictions about the thermodynamic properties of numerous physical systems, predictions which no other theory can make. Georgesçu-Roegen's contention that statistical mechanics had no experimental verification (*The Entropy Law and the Economic Process*, 1971: 161) is far from the truth today.

Another controversy concerns the twentieth-century idea, due to Shannon and Weaver, that thermodynamic entropy is related to information content, and therefore that the Second Law of Thermodynamics says something about information or knowledge. In its most extreme form, advocates of this approach take the position that scarcity of low-entropy natural resources can be outweighed by an abundance of human knowledge. Georgesçu-Roegen raised strong objections to any connection between the Second Law and information, writing: "How our ignorance – a subjective element – can be a coordinate of a physical phenomena, like that expressed by the Entropy Law, is beyond the sheerest fantasy" (*The Entropy Law and the Economic Process*, 1971: 162–3; see also his entire Appendix B). Georgesçu-Roegen's correct position is supported by Mayumi (1997), Wicken (1987), and a long host of authors cited by Lozada (1999). Wicken's argument is particularly careful and insightful without being technical. Below we give an extensive excerpt of it (1987: selections from 19–27); Georgesçu-Roegen would no doubt have been pleased to have such an eloquent comrade:

Willard Gibbs was able to show that

$$S = -k\Sigma P_i \ln P_i \qquad (1-3)$$

where P_i refers to the energy-dependent probabilities of the various microstates . . .
The Shannon equation is

$$H = -K\Sigma P_i \log_2 P_i \qquad (1-4)$$

where K is generally taken as unity. Since proportionality constants and logarithm bases are more matters of convenience and scaling than of substance, the relationships among the variables in the two equations are identical. Gibbs circumspectly referred to his statistical formulations as "entropy analogues" rather than "entropies" (Denbigh 1982). The question is then whether the Shannon equation generalizes the *entropy analogues* of statistical mechanics.

The answer to this question depends on whether Shannon entropies have properties consistent with thermodynamic entropies. Issues relevant to it are: (a) do both entropies behave the same way? and (b) are they both based on the same *kinds* of probabilistic assumptions? Neither is the case.

The generalization claim would automatically be true if concepts were reducible to equations. They are not. Terms in equations are defined according to context of application, and derive their meanings accordingly. Boltzmann did not invent the equation on which his statistical treatment of entropy was based . . . Boltzmann found a new application for it in thermodynamics. Shannon yet another in communications. The generalization question then hangs on the natures of the terms involved in the respective equations. These involve uncertainties and probabilities.

The uncertainties involved in statistical thermodynamics and information theory are of very different natures. In statistical thermodynamics, that uncertainty is *fundamental*: one cannot know the microstate (or N-particle quantum state) of a physicochemical system precisely because it doesn't reside in any specific microstate, but fluctuates stochastically through a set of alternatives . . .

In information theory, uncertainty is strictly the before-the-fact variety involved in the specification of messages from alternative sequences of symbols . . .

The macrostate/microstate distinction on which the Boltzmann–Gibbs equations are based does not apply to the Shannon entropy. Since the explanation of entropic increase in irreversible processes is *dependent* on that distinction, both questions in the generalization must be answered negatively. While the Shannon equation is symbolically isomorphic with the Boltzmann equation, the meanings of the symbols in the respective equations have little in common . . .

Subscription to the generalization thesis mars Gatlin's (1972) otherwise lucid introduction to the fundamentals of information theory and its biological applications . . . [Here is reproduced a paragraph from Gatlin about the entropy of different arrangements of Scrabble pieces.] This paragraph [of Gatlin's] reveals the enormously broad connotative field that has historically enshrouded the entropy concept since its statistical treatment. Entropy has become associated with disorder in all its manifestations and order has in turn been equated to organization. These are very loose semantics . . .

As a result of its independent lines of development in thermodynamics and communications theory there are in science today two "entropies." This is one too many. It is not science's habit to affix the same name to different concepts. Shared names suggest shared meanings, and the connotative field of the old tends inevitably to intrude on the denotative terrain of the new.

"Entropy" was not a term than Shannon had himself decided on by simple virtue of the formal similarity of his equation to Boltzmann's. As Denbigh (1982) discusses, he did so at the urging of von Neumann for very practical reasons. Entropy was an established concept that his theory could draw on for salability to the scientific community. Science was the loser for this choice.

Wittgenstein (1973) once defined the job of philosophy as the "battle against the bewitchment of our intelligence by means of language." The casual usage of such terms as "disorder" and "disorganization" to provide pictorial clarity to the concept of entropy has served primarily to obscure its meaning. To the case at hand: there is an inevitable tendency for connotations to flow from the established to the new, and the Shannon entropy began from the beginning to take on colorations of thermodynamic entropy . . .

Since the equation $H = -K\Sigma P_i \log P_i$ has such a rich variety of applications – from games of chance to statistical thermodynamics to complexities of structures – it should be treated in the manner of a standard algebraic equation, with H named according to context of application. In statistical mechanics, "entropy analogue" is appropriate. Entropy measures a lack of fixed structured relationships or fixed energy assignments. Where there is structure, it should be acknowledged from the beginning by abandoning "entropy', in favor of "complexity." There is no need in information theory to enter the "entropy means lack of information" arena, or even to talk about entropy . . .

To appreciate the importance of restricting entropy to thermodynamic applications – or, more broadly, to applications in which a macrostate–microstate relationship obtains – one need only reflect on Weaver's remarks about the Shannon formulation making contact with a universal law [the Second Law of Thermodynamics]. It does no such thing . . .

If it were possible to treat "entropy" simply as an equation, with properties dependent on area of application, calling Shannon's function by that name would be relatively unproblematic. However, most who use the term "entropy" feel something of Weaver's conviction about contacting a universal principle that provides sweeping laws of directional change. Precision in the use of terms is an important mechanism for keeping Spencerian ambitions in check. (parentheses added)

None of this is to deny that the quantity called "entropy" by communication theorists or statisticians has usefulness in those fields. The point is that "entropies" in contexts where temperature is absent have nothing to do with the entropy of thermodynamics and nothing to do with the Second Law of Thermodynamics. (See Morowitz, 1991: 235–6.)

In summary, Georgesçu-Roegen's understanding of entropy was, with a few exceptions, solid. He saw the errors inherent in thinking that the Second Law was relevant to information, in believing in Boltzmann's H-theorem, in thinking that entropy was cardinal rather than weakly cardinal, and in thinking that energy is the answer to all resource problems.

6.6   ENTROPY AND ECONOMIC EPISTEMOLOGY AND METHODOLOGY

Throughout his career, Georgesçu-Roegen was interested in phenomena which were novel and whose emergence could not be predicted even in principle. Frank Knight used the term "uncertainty," as opposed to risk, to discuss such phenomena (*The Entropy Law and the Economic Process*, 1971: 122). Georgesçu-Roegen opposes terms such as change, history, irreversibility, quality, novelty, true happening, and evolution to terms such as locomotion, ahistorical, reversible, qualityless, mechanism, and geometry. He is interested in situations when consumers choose a bundle X thinking it will give a utility U(X), only to be shocked when they discover the utility level is different from what they ever could have imagined. (The subsequent permanent change in their utility function is an example of "hysteresis.") Analogous things happen to firms.

A convincing argument that economists should pay attention to genuine novelty can be developed without mentioning anything about physics. In *The Entropy Law and the Economic Process* Georgesçu-Roegen did not try. He wrote that by the time of Jevons and Walras:

> a spectacular revolution in physics had already brought the downfall of the mechanistic dogma . . . None of the latter-day model builders seem to have been aware at any time of this downfall. Otherwise, one could not understand why they have clung to the mechanistic framework with the fervor with which they have. (*The Entropy Law and the Economic Process*, 1971: 3)

The last two sentences of this quotation explain why Georgesçu-Roegen used arguments from physics rather than from economics when suggesting what epistemology economists should use. In retrospect, this was a tactical mistake. Georgesçu-Roegen did not realize that to most of his fellow twentieth-century economists, physics was epistemologically irrelevant.

It is therefore not surprising that economists have seen no epistemological relevance to insights such as "the Entropy Law . . . is the simplest form by which the existence of true happening in nature is recognized" (ibid.: 169), or "the Entropy Law . . . marks the recognition by . . . the most trusted of all sciences of nature that there is qualitative change in the universe" (ibid.: 9–10). Georgesçu-Roegen's discussion of the emergence of novelty in biological evolution or in the joining of chemical compounds, and his demonstration that one of the premier physical laws, the Second Law of Thermodynamics, is an evolutionary law (ibid.: 128–9) evoked from economists no response. Instead of providing so many "object lessons from physics," Georgesçu-Roegen would have been better received had he provided more object lessons from economics.

Changing now from epistemology to methodology, we begin by taking note of Georgesçu-Roegen's high praise for the use of arithmomorphic techniques in economics (ibid.: 35, 50, 337). He wrote: "Whenever arithmetization can be worked out, its merits are above all words of praise" (ibid.: 15), and "the human mind can comprehend a phenomenon clearly only if it can represent that phenomenon by a *mechanical model*" (ibid.: 141). He even wrote that "it is ghastly to imagine the destruction of all present capital equipment, still ghastlier to think of all men suddenly forgetting all mathematics" (ibid.: 332).

However, Georgesçu-Roegen felt that arithmomorphic models could not describe all phenomena of economic importance. Even in the physical world, while the Entropy Law sometimes gives definite numerical predictions about phenomena, as we have shown in section 5.5,[14] in other situations it only shows the arrow of time, without giving a particular arithmomorphic rate or other number. Evolutionary relationships cannot be captured using the same arithmomorphic methodology as other relationships, because "there is an irreducible incompatibility between qualitative change, i.e., between essential novelty, and arithmomorphic structures" (*Energy and Economic Myths*, 1976: 237). He expanded:

In principle, we can indeed write the equations of any given production or consumption process (if not in all technical details at least in a global form). Next, we may either assemble all these equations into a gigantic system or aggregate them into a more manageable one. But to write any set of the initial equations, we must know the exact nature of the individual process to which it refers. And the rub is that in the long run or even in the not too long run the economic (as well as the biological) process is inevitably dominated by a qualitative change which cannot be known in advance. Life must rely on novel mutations, if it is to continue its existence in an environment which it changes continuously and irrevocably. So no system of equations can describe the development of an evolutionary process. (*The Entropy Law and the Economic Process*, 1971: 17)

In other words, arithmomorphic models are incapable of analyzing genuinely novel change, so if an economist accepts Georgesçu-Roegen's epistemological position that genuinely novel change is an important characteristic of economies, he had better have more than arithmomorphic models in his toolbox.

These epistemological and methodological positions create a large gap between Georgesçu-Roegen's way of doing economics and the way most neoclassical economists do economics.

For example, since Georgesçu-Roegen felt novelty was an important characteristic of economic development, he frowned upon "linear thinking" (*Energy and Economic Myths*, 1976: 14), the simple extrapolation of the past to the future. This was his point in quoting "the famous old French quatrain: 'Seigneur de La Palice / fell in the battle for Pavia. / A quarter of an hour before his death / he was still alive'" (ibid.: 16 fn. 30). Georgesçu-Roegen felt neoclassical economists were merely extrapolating the past few centuries' "mineralogical bonanza" into the future to arrive at their "Panglossian" or "cornucopian" predictions that "come what may, we shall find a way," that nothing "could ever stand in the way of an increasingly happier existence of the human species"; he complained: "it is utterly inept to argue . . . that since mankind has not met with any ecological difficulty since the age of Pericles, it will never meet with one."[15] The alternative is to build the Second Law constraints into the model. However, the Second Law minimum energy constraint concerns energy in its physical sense, unrelated to a specific class of economic commodities. Furthermore, that constraint is strict, whereas mathematical optima can only be guaranteed to exist when inequalities are weak. (Imagine minimizing x under the constraint that $x > 0$.) For these two reasons, it is not easy to maintain a neoclassical framework and simply incorporate Second Law considerations with a new constraint.

To be more concrete, we have to specify what the neoclassical framework for resource economics is. If t is time, r is the subjective rate of discount, x is the stock of a natural resource, q is the flow of resource, and $F(x)$ is the rate at which the resource regenerates ($F \equiv 0$ for an exhaustible resource), then consider the problem of maximizing over (say) all piecewise continuous functions q_t, the quantity $\int_0^\infty e^{-rt} A(q_t)\, dt$ subject to $q_t = -dx/dt$, $x_0 > 0$, and $q_t \geq 0$ and $x_t \geq 0$ for all t. If A is social welfare, this is a neoclassical social-welfare-maximizing extraction problem. If A is profit, this is the problem faced by a neoclassical profit-maximizing firm, as was addressed by Hotelling (1931) with $F \equiv 0$. There is a large literature on models of this sort, where variations include non-constant r, x_0 being a random variable, F having a random component, imperfect competition, more than one deposit of the resource, non-concave A, and so on. Georgeşçu-Roegen's main objection to this kind of model, when used to find a socially optimal extraction path, was that it assumes a strictly positive r, which means the welfare of people in the future is counted less than that of people today:

> For quasi-immortal entities – such as a nation and especially mankind – discounting the future is wrong from any viewpoint . . . if all future utilities are treated alike, the beautiful solution reached by Hotelling is of no use anymore . . . The *analytical* solution is to spread resources evenly in time, which in the case of an infinite time horizon yields the paradoxical result that each year a null amount of resources should be consumed. ("Comments on the papers by Daly and Stiglitz," 1979: 101–2)[16]

There are neoclassical models which use $r = 0$, such as Solow (1974) and Toman *et al.* (1995); Georgeşçu-Roegen objected to their specification of technology (see below).

Seen from a general equilibrium point of view, the interest rate is an endogenous price determined by market supply and demand, not an exogenous parameter amenable to change, nor one of the exogenous individual subjective discount rates. (Exogenous individual subjective discount rates may exist in such a model, but they are different from the endogenous market-clearing rate, and any difference between the individual rates and the market-clearing rate induces some people to be borrowers and others to be lenders.) In such a framework, the welfare of each generation is determined by its initial endowment of goods (together with its technology). If generations did not overlap, future generations would have to depend entirely on the present generation's altruism for their well-being. When generations overlap this is less the case, but it is still true that the welfare of each generation depends on its initial endowment, plus any intergenerational transfers effected by individuals or the government for altruistic or other purposes. As Herman Daly (1993: 329) has written: "intertemporal distribution is a question of ethics, not a function of the interest rate." Howarth and Norgaard (1995) have written a wonderful overview of overlapping-generations models like these.

Georgesçu-Roegen, though, was interested in realistic descriptions of technology, and overlapping-generations models should not be expected to have gone beyond received theory in that respect. "Received theory" means that the inputs to production, mainly resource R, labor L, capital goods K, and perhaps pollution X, combine to create "stuff"[17] according to an aggregate production function $f(R, L, K, X)$, which is often assumed to take the Cobb–Douglas form $f = R^\alpha L^\beta K^\gamma X^\delta$ where all the Greek letters but δ are positive. "Stuff" then gets divided into consumption C and investment dK / dt.

The Cobb–Douglas form has a property which infuriated Georgesçu-Roegen: f may be kept constant even if $R \to 0$ as long as there is enough L and K to substitute for the dwindling R.[18] Georgesçu-Roegen (*Energy and Economic Myths*, 1976: 17) wrote that such an assumption "ignore[s] the difference between the actual world and the Garden of Eden." He would have preferred a model in which as R goes to zero, K would have to go to zero. Neoclassicals have written about this case (Dasgupta and Heal, 1979: 197–8), but have seemed mostly bored by it because it implies that $\lim_{t \to \infty} f = 0$ so trivially. Our position is that constraints are what makes economics interesting, and that even if humanity is doomed to eventually die off, there are important resource-allocation questions to answer until humanity reaches that point. Daly (1993: 330) has taught that "growth" and "development" are different, and one can have the latter even in an environment not allowing the former.[19]

Georgesçu-Roegen would have found many more economists interested in his work if he had chosen to improve on the Cobb–Douglas formulation instead of merely criticize it. Certainly, the Cobb–Douglas formulation is wrong, but all models used in theoretical economics are "wrong" in some sense, and that is what makes them useful: as Varian (1996: 1–2) points out, an economic model which is right in every detail is as useless as a map drawn on a scale of one to one. What is most likely, however, is that Georgesçu-Roegen failed to improve on these models not because he preferred the role of a critic, but rather because he had become convinced that arithmomorphic models of any sort were unsuitable for discussing the future, for the fundamental reasons cited earlier in this section: historical developments have to be evolutionary, and evolutionary developments cannot be described by systems of equations. As Georgesçu-Roegen wrote ("Comments on the papers by Daly and Stiglitz,"

1979: 98): "the question that confronts us today is whether we are going to discover *new* sources of energy that can be safely used. No elasticities of some Cobb–Douglas function can help us to answer it."

This position poses a considerable challenge to the way we do economics. It certainly challenged Georgesçu-Roegen. He wrote: "whenever we may try to prescribe a quantitative policy for the economy of resources we can only play the tune by ear" (ibid.: 102). In another passage (*Energy and Economic Myths*, 1976: 19, emphasis added), he said:

> Sir Macfarland Burnet, a Nobelite, in a special lecture considered it imperative "to prevent the progressive destruction of the earth's irreplaceable resources . . . ." And a prestigious institution such as the United Nations, in its Declaration on the Human Environment (Stockholm, 1972), repeatedly urged everyone "to improve the environment." Both urgings reflect the fallacy that man can reverse the march of entropy. The truth, however unpleasant, is that the most we can do is to prevent any unnecessary depletion of resources and any unnecessary deterioration of the environment, but *without claiming that we know the precise meaning of "unnecessary" in this context.*

Georgesçu-Roegen (*Analytical Economics*, 1966: 274–5) felt arithmomorphic techniques failed not only for planning the far future but even in the business of everyday life:

> The difficulty of formulating an objective criterion for decisions involving Knightian uncertainty stems precisely from the fact that the corresponding expectations are not ordinally measurable, probably not even comparable, and have only a distant and debatable connection with past observations. Under such circumstances there seems to be no other recommendation for dealing with Knightian uncertainty than the common advice: "get all the facts and use good judgment." But what is "good judgment"? The concept seems to resist any attempt at an objective definition that also would be operational *ex ante*. . . . Together with gathering, presenting, and analyzing in a logical fashion as many facts as possible, to detect and to use good judgment constitute the only means by which we can respond to living without divine knowledge in an uncertain world. To many this may sound very discouraging, but the opposite view, that good judgment is an obsolete concept in an era of panlogistic models, is patently delusive.

So, what is an economist to do? First, undertake historical and institutional studies, as well as arithmomorphic ones.[20] Then, make proposals. Georgesçu-Roegen did this, forwarding at least general proposals both for the conduct of public policy and for the conduct of individual citizens in light of the entropic degradation of the earth. These proposals form part of the subject matter of the next chapter.

NOTES

1. Entropy increases because the duck dies. Death occurs when a living organism is put in many kinds of isolated systems (for example, one without a source of food); since entropy only increases in such systems, death increases entropy. Alternatively, one can argue that living cells import and expend energy to keep some otherwise-spontaneous chemical changes (including some chemical mixings) from occurring. Once death occurs, the energy importation stops and the spontaneous changes occur. These spontaneous changes must increase entropy because the system approximates being an isolated system just after death, and in an isolated system, the changes which decrease entropy cannot occur.

2. Of course no process solely decreases entropy; the entropy of the universe always increases.

3. Georgesçu-Roegen wrote that the maximum amount of mechanical work that can be derived from a system is given by F. He called what is left of U beyond F, namely TS, unavailable ("bound") energy. If U and T could be taken as constants, and if the system could be isolated so its entropy was increasing, then as time went on, F would fall. Georgesçu-Roegen concludes that the Entropy Law implies "that the relative importance of unavailable energy of an isothermal system continuously and irrevocably increases." Compare Georgesçu-Roegen's statement with this paraphrase of the one in Fermi (1937/1956: 78–9):

   > If a system suffers a reversible (irreversible) transformation from an initial state A to a final state B both of which states have a temperature equal to that of the environment, and if the system exchanges heat with the environment only, then during the transformation, the work done by the system is equal to (is less than) the decrease in the Helmholtz free energy of the system.

   Fermi's statement shows that F is a measure of the work that can be done by a *non-isolated* system; such a system need not have increasing S, and so even if U and T are constant, F need not be falling. Furthermore, since the system exchanges heat with its environment, U may not be constant; thus even if S did rise for a period of time, F may not be falling then. In addition, the restriction to constant T means that even if F did fall, the interpretation of F as maximum potential for work is not applicable to systems which have temperature differences or which have interactions with an environment having a different temperature. Examples of such systems are the entire universe, which is not isothermal, and the Earth, which is warmer than outer space, with which it exchanges energy. In summary, there is no broadly applicable law of falling potential for work based on ever-increasing S diminishing the universe's stock of $F = U - TS$.

4. The energy in a bucket of room-temperature uranium is of no use to me if I stay inside the room and have no knowledge of atomic energy, but if I can hook a heat engine up between the bucket and cold winter air out of doors, then some of this energy is useful, and if I have a nuclear reactor handy (one which society allows to operate), then even more of it is useful. What is the amount of non-degraded energy in the bucket? To show what a difficult question this is, consider that in trying to determine whether life will be possible forever despite the Entropy Law, the astrophysicist Frautschi (1982) contends that an extremely important potential source of low entropy in the far future will be the bringing together of black holes. Frautschi concludes that living solids cannot exist forever, but other forms of life might be able to. (See also Goldstein and Goldstein, 1993: 388.)

   Zemansky (1968: 80) points out another difficulty with trying to measure the "amount of non-degraded heat energy," writing: "Heat is energy in motion . . . It would be just as incorrect to refer to the 'heat in a body' as it would be to speak of the 'work in a body.'" Shifting the focus from heat to internal energy U does not help because U is weakly cardinal (Fermi, 1937/1956: 13–14).

5. After all, Georgesçu-Roegen wrote ("Thermodynamics and we, the humans," 1991: 199) that "The life spans of species are ordinarily long and ours should not be exceptional in that respect. It is from this eonic viewpoint that we should base any programmes for maximizing Our survival."

6. On that page he wrote: "In an environment of very low entropy, a living organism would not be able to resist the onslaught of the free energy hitting it from all parts." Actually, a low-entropy environment (if "low" could be precisely defined) need not entail any free energy hitting anything, it would be a collection of "low"-entropy matter.

7. See also the last part of this: "the Entropy Law is the taproot of economic scarcity. Were it not for this law, we could use the energy of a piece of coal over and over again . . . Also, engines, homes, and even living organisms . . . would never wear out. There would be no economic difference between material goods and Ricardian land. In such an imaginary, purely mechanical world, there would be no true scarcity of energy and materials" (*Energy and Economic Myths*, 1976: 9).

8. Georgescu-Roegen ("Matter matters, too," 1977: 300ff) knew about both (1) and (3); what got him into trouble was that he did not know that (3) could be derived from (1). The proof is not easy, the result is not obvious, and the consequences of not understanding the argument are severe, so all the details were provided at the start of section 5.3.

9. "Ever since my first thought on the entropic nature of the economic process, my position has been that in any system that performs work of any kind not only free energy but also '*matter arranged* in some definite structures' continuously and irrevocably dissipates . . . [I have since discovered that] the energetic dogma constitutes the general view. I then began to offer specific arguments to prove that matter matters, too, which led me to formulate a new law, to which I referred as the fourth law of thermodynamics (a not very fortunate choice)" ("Energy, matter. and economic valuation," 1981: 58–9).

10. Another example is: "The amount of accessible energetic low entropy is finite . . . the amount of accessible material low entropy is finite, too" (*Energy and Economic Myths*, 1976: 11), which shows both the entropy/energy difficulties and the weakly cardinal/cardinal difficulties addressed in section 6.2.

11. Chapman and Roberts (1983: 97) write: "The goal of an energy analysis is to evaluate the total quantity of energy" then add the following footnote: "This is the energy content of the fuels used. It does not include energy (in the thermodynamic sense) extracted from the environment or from non-fuel materials (such as sulfur). Hence it was once pointed out . . . that a better name for this activity would be fuel analysis or even fossil fuel accounting." The rub is that the Second Law of Thermodynamics, on which Chapman and Roberts base their argument that "energy is special," only concerns "energy in the thermodynamic sense."

12. Statistical mechanics does not define absolute entropy, even taking the Third Law of Thermodynamics into account (Dugdale, 1996: 99, 100, 146). We believe the Planck/Zemansky mathematical argument's error is to start by postulating that systems have "entropy," whereas systems actually only have entropy differences between their current state and a reference state. Specifically, Planck starts by assuming that S is some function of the "thermodynamic probability" W, so $S = f(W)$, then argues that the entropy of a composite system of objects '1' and '2' would be $S_1 + S_2 = f(W_1) + f(W_2)$, while it would also be $f(W_{12}) = f(W_1 W_2)$. It follows that $f(W_1 W_2) = f(W_1) + f(W_2)$. Operating on both sides with $\partial^2 / \partial W_1 \partial W_2$ and letting $W = W_1 W_2$ yield $f'(W) + W f''(W) = 0$, whose solution is $f(W) = k \ln W + C$. However, C = 0 in order to satisfy $f(W_1 W_2) = f(W_1) + f(W_2)$. Therefore $f = k \ln W$ and $S = f$. We claim that the correct derivation is to start by assuming that $S - S_0 = f(W)$. The entropy of a composite system would be $(S_1 - S_0) + (S_2 - S_0) = f(W_1) + f(W_2)$, while it would also be $f(W_{12}) = f(W_1 W_2)$. Then $f(W_1 W_2) = f(W_1) + f(W_2)$, so as before $f = k \ln W$, only this time the implication is $S - S_0 = f$, leading to the correct $S = k \ln W + S_0$ ((5.4)), not to Boltzmann's and Planck's S = $k \ln W$. Especially with the Third Law of Thermodynamics, one can without loss of generality take $S_0 = 0$, so operationally the distinction between (5.4) and $S = k \ln W$ is not important. The distinction is important in theory because it is the difference between entropy being cardinal or weakly cardinal. Plank's mathematical argument given above for why the differential equation can only be solved by setting C = 0 comes from p. 118 of the fifth German edition; the argument given on p. 119 of Masius' translation of the second edition is unsatisfactory.

13. A detailed treatment of this topic, including discussion of the current controversy between Ilya Prigogine and the "Astrophysical School" concerning the origin of the arrow of time which the Second Law describes, is contained in Lozada (1995). A more recent, technical treatment

is given by Schulman (1997: esp. §4.1); while we do not agree with all the ways Schulman defines entropy, he clearly rejects the H-theorem origin of the arrow of time and embraces the cosmological origin.

14. This does contradict Georgesçu-Roegen (*Energy and Economic Myths*, 1976: 9): "Equally important is the fact that the Entropy Law is the only natural law that does not predict quantitatively."

15. The sources are as follows (Georgesçu-Roegen is the author of all): "mineralogical bonanza" ("Comments on the papers by Daly and Stiglitz," 1979: 97); "Panglossian" ("Thermodynamics and we, the humans," 1991: 193); "cornucopian" ("Looking back," 1991: 20); "come what may. we shall find a way" (*Energy and Economic Myths*, 1976: xv); the second-to-last quote (*Energy and Economic Myths*, 1976: 16); the last quote (ibid.: 5).

16. Formally: for strictly concave A, with $r = 0$ a plan which extracts x_0 / T of the resource for T time periods can always be improved upon by another plan which picks a larger value for T; however, the limit of such plans is the plan which extracts nothing forever, which minimizes rather than maximizes the objective function over all feasible plans. See also Artstein (1980).

17. The term comes from Bliss (1975: 281). Joan Robinson called it "leets", which is "steel" spelled backwards (Burmeister, 1980: 138).

18. See Daly (1997) and Georgesçu-Roegen (*Energy and Economic Myths*, 1976: 17): "*substitution within a finite stock of accessible low entropy* whose irrevocable degradation is speeded up through use cannot possibly go on forever."

19. Georgesçu-Roegen (ibid.: 19) says the distinction between growth and development is due to Schumpeter.

20. "I would be among the last servants of science to deny the indispensable role of theory, which must necessarily aspire to be quantitative and hence mathematical, provided 'theory' is not separated completely from fact. But, as my master Joseph A. Schumpeter did so poignantly, I would also be among the first to defend the absolute necessity of historical and institutional studies in social sciences, hence in economics" (*Energy and Economic Myths*, 1976: xi).

7. Bioeconomics

7.1 INTRODUCTION

Bioeconomics represents the culmination of Nicholas Georgesçu-Roegen's work in economics and thermodynamics. Although the term "bioeconomics" was not coined by him, Georgesçu deserves the credit for formulating bioeconomics as credible academic endeavor.[1] Unfortunately, he never completed an advertised book on this subject, so his views must be inferred from a wide collection of his later writings. However, unlike most of his methodological work, Georgesçu's expositions of the principles of bioeconomics are relatively consistent through time. We are thus able to summarize his views with some confidence.

The origins of Georgesçu's bioeconomic ideas are interesting. In his auto-biographical essay in Szenberg (1992), he credits two factors with his awakening awareness of Man's bioeconomic predicament: Emile Borel's writings on thermodynamics (which Georgesçu read in the 1920s), and his experience of living in Romania in the 1920s and 1930s. Further, Georgesçu says: "the thought that, even if the population stopped growing, its predicament would still remain came to me one day as I watched a big Romanian river running in its bed furiously and with a chocolate color. There goes, I said, our daily bread of tomorrow."[2]

The basic tenets of bioeconomics are, to a remarkable degree, consonant with longstanding themes in Georgesçu's work. For example, his earlier work on utility theory utilized the notions of hysteresis, evolution, and the inadequacy of the real numbers for representing certain phenomena.[3] These notions are fully concordant with the representation of economic activity adopted by bio-economics. The basic principles of bioeconomics are:

1. Economic activity is an aspect of man's biological existence.
2. Economic activity, and all life processes, requires a steady diet of energy and "low entropy" material from the environment.
3. All economic activities necessarily represent a deficit in entropy terms, and result in a reduction in the stock of useful matter available to sustain the human race.
4. There is a "dictatorship of the present over the future".

5. Humans, unlike animals, use created "exosomatic" tools to gather low entropy, and this circumstance leads to social conflicts.
6. With the exceptions of animal husbandry, fire and heat engines, denoted "Promethean" technologies, technology is "parasitic" and cannot form the basis of a viable social system.
7. The bioeconomic problem of mankind can be solved only through changes in human values, and such changes would need to be accompanied by significant alterations in politics and economic behavior.

As a review of these "axioms" makes clear, bioeconomics is both a methodology for doing economics and a set of overarching recommendations for the proper conduct of human affairs. The methodological aspect demands that we view economic activity within the context of the natural world, a request no one can properly dispute, although opinions differ as to the ecological damage imposed by human activities. The *basis* for the claim that economic activity is dependent on the natural world is thermodynamics (for example, the minimum energy requirements for transforming any desirable material imposed by the Entropy Law), combined with the observation, discussed in a previous chapter, that living things sustain themselves on a flow of "low-entropy" matter (or energy used to refine matter) from the environment. It is also true that human economic activity "degrades" matter into "waste" in the sense that utilization of materials results in their dispersion throughout the environment. In Daly's often quoted example, using automobile tires causes the rubber particles of which they are made to disperse widely into the environment.[4] Even if one could reassemble them in principle, the energy requirements to do so would clearly be astronomical.[5]

In order for matter to "matter" in a critical sense, some limits on recycling must exist (else we could, in Georgesçu's phrase, "all eat the same dinner"). Such limits could be either (a) absolute limits, such as that imposed by the Fourth Law; (b) practical limits imposed by our technological inability to access energy, or (c) limits imposed by the unavailability of sufficient energy itself.

Circumstances are worse than this description implies, however, because recycling itself requires capital equipment made of materials, which are in turn refined and made by equipment made of materials, and so on and so forth. One is reminded of Georgesçu's observation that, barring technical change, it is unclear how an economy described by Leontief's system is supposed to grow. When everything depends on everything else, an increase in one flow requires a "prior" increase in (at least) one other flow, which requires in turn a similar increase in another, and so on.[6]

The Fourth Law, then, represents an important idea in Georgesçu's bioeconomic vision. This law, however, is not essential for his purpose in general: both limits on energy and its accessibility to us imply that "matter matters." However, arguments of this latter sort are intellectually less satisfying than invocations of a "Law." Although no one has conclusively demonstrated that the Fourth Law is true (or false), it is critical for part of our exposition of bioeconomics *à la* Georgesçu–Roegen that we assume that the Fourth Law is "practically true," that is, economic activity does result in material dissipation, and that this general dissipation will not be completely reversed by technological means.[7] Given this background, Georgesçu–Roegen's explanation of the "social conflict" may be described.

7.2   BIOECONOMICS, THE STRUGGLE FOR "LOW ENTROPY," AND SOCIAL CONFLICT

Animals, including *homo economicus*, are completely dependent for their continued existence on propitiously distributed, highly ordered material structures. Although plants use energy, such as solar radiation, to create ordered structures, people and animals are not able to maintain themselves without consuming ordered, "low-entropy" materials. The natural environment is the source of these materials, and both humans and the lower animals use instruments to gather beneficial ordered materials from the environment. Although, as Georgesçu acknowledges often, Man cannot be defined as the (sole) "user of tools," Man does appear to be the only "maker of tools by means of tools." Animals use "endosomatic organs," that is, their limbs (which belong to them from birth), to gather ordered materials, while humans use "exosomatic organs", such as shovels and shotguns, to harvest nature's bounty. Unlike the limbs or organs of animals, though, the distribution of exosomatic organs is not naturally ordained, but instead reflects *social* forces. Because possession of these instruments may convey great benefits, rivalries emerge for their control. Georgesçu viewed such rivalries as the basis for Man's social conflict.[8] Only men can have "civil wars." As Georgesçu argues in *The Entropy Law and the Economic Process* (1971), human nature, which allows us to theorize, create art, and understand the natural world, also allows the peasant farmer, or the "coolie," to imagine her- or himself as the Brahmin.[9] Because the allocation of exosomatic tools is itself a human contrivance, men have long expended themselves in rearranging the claims these tools entail. Social conflict and war is the result.

Georgesçu's view that the exosomatic nature of Man's tools is the basis of social conflict might be dismissed by some observers as a commonplace. It is important for understanding his bioeconomic vision, however, that this should not occur. First, Georgesçu attached great weight to this idea, and spoke or wrote about it very often.[10] Second, it must be remembered that Georgesçu-

Roegen was the product of an intellectual environment in which Marx was taken very seriously. Although it is quite hazardous to summarize Marx, many readers would agree that Marx believed the transformation of society to communism would "solve" the social conflict arising from the distribution of ownership of the "means of production."[11] Against this backdrop, it is apparent that Georgesçu's explanation for the social conflict may be a reaction to Marx's more optimistic scenario. In Georgesçu's view, the social conflict arises from the combination of (a) biological necessity; (b) limited "low-entropy" materials; (c) Man's essential nature. While factors (a) and (b) are presumably unalterable, could not Man's nature be beneficially changed?

Georgesçu devoted considerable space in *The Entropy Law and the Economic Process* to a closely reasoned condemnation of the "perfectability" of men through scientific manipulation of genes, and to other "utopian" proposals for the ordering of society.[12] He here illustrates again his doubts that "technology" will be Man's salvation. Man can change, but it was pure fantasy, in Georgesçu's view, to imagine that technological means would be found to rescue humans from their worst impulses. His views are similar to those of C.S. Lewis in "The abolition of man", reprinted in Daly and Townsend (1993). Georgesçu-Roegen also doubted that appeals to moral sentiments would change general human behavior. He wrote:

> Will mankind listen to any program that implies a constriction of its addiction to exosomatic comfort? Perhaps the destiny of man is to have a short but fiery, exciting, and extravagant life rather than a long, uneventful, and vegetative existence. Let other species – the amoebas, for example – which have no spiritual ambitions inherit an earth still bathed in plenty of sunshine.[13]

Conflicts arising from attempts by humans to redistribute exosomatic tools will include both international and civil wars. Georgesçu saw both types of violence as essentially similar. His explanation for civil wars is especially illuminating, as it illustrates both the consistency between some of his sociological views and those of Marx, and his willingness to integrate economics, biology, physics, and sociology into a whole. Georgesçu argued that the use of exosomatic tools requires the emergence of "supervisory classes," that is, managers and bureaucrats, whose contributions to output are difficult to measure. As a consequence, the "values" of their services can be safely exaggerated, leading, according to Georgesçu, to a recurring characteristic of the elites throughout history: elites are composed of persons whose productivity has no objective measurement. Bricklayers, fishermen, and so on can never constitute the elite, while priests, astrologers, and bureaucrats luxuriate in every age.[14]

As with others of Georgesçu-Roegen's propositions, these claims (on the sources of the supervisory classes' wealth) are both interesting and confusing. His idea is not precisely articulated, and it is perhaps useful to attempt to recast it in more modern economic language. The modern theory of principal and agent, and the analysis of contracts provide such a language.

As a general proposition, the inability of one party (say, the principal) to accurately measure the effort (contribution to expected output) of another party (the agent) ordinarily leads to contract provisions that reflect this imperfection in monitoring.[15] Georgesçu's notion seems to be that this inability leads to "excessive" compensation for the supervising class. Clearly, no such conclusion is generally warranted solely by the inability of one party to monitor the effort of another. When, as the simplest principal–agent model assumes, the principal has all of the bargaining power, the agent may receive only a minimally favorable outcome. Thus, in order for Georgesçu's proposition to be correctly evaluated, one should recognize the role of bargaining power and, perhaps, other sorts of asymmetric information generally. As is usual with economic propositions containing a notion of social classes, neoclassical economists are likely to find Georgesçu's views in this area unconvincing.

Perhaps one should take his explanation of class conflicts and civil wars, including the "war" between the town and countryside (discussed in Chapter 4), as empirical generalizations, rather than analytic propositions derived from some (unrevealed) model. At least one of his former students described Georgesçu as "the world's greatest casual empiricist." Certainly the observation that supervisory classes grow rich while exaggerating their contributions is hard to deny. Georgesçu-Roegen's first-hand experiences in government service in Romania no doubt bolstered his conclusions.

7.3   TECHNOLOGIES, PROMETHEAN AND OTHERWISE

Georgesçu's long interest in the economics of production, as represented both by his earlier work on Leontief models and his later development of the flow-fund model, greatly influenced his writings in bioeconomics . When Georgesçu defines economics itself as "the study of transformations in matter and energy brought about by human action and entropy," the centrality of the notion of production is apparent. "Production" played several roles in his bioeconomic program, and it is the purpose of this section to examine them.

Many of Georgesçu-Roegen's discussions of bioeconomics make use of an important distinction among "technologies" – with that term broadly conceived; this distinction arose naturally from his emphasis on the notion of "boundary" in the definition of a process.[16] Georgesçu asks us to consider "technology" as a term describing human economic activity as a whole, and to examine the predominant characteristics of this activity over the millennia. In this framework, he designated a technology as "Promethean" if it is simultaneously self-supporting and capable of being the basis of an ongoing social system. In

Georgesçu's sense, a technology is "self-supporting" if it harvests sufficient resources to fund its own contrivance (if only indirectly), with something left over. For example, the technology based on fire is termed Promethean because, in Georgesçu's words, "with just the spark of a match we can set on fire a whole forest, nay, all forests."[17] The only other Promethean technologies ever cited by Georgesçu are animal husbandry and heat engines.

Some insight into precisely what he intends is offered by his further distinction between "available" resources and "accessible" resources. To highlight this, Georgesçu remarks:

> Let us assume that another earth would possess vast reserves of bituminous coal located 10^7 feet below the surface. Since it would take more than the energy of one pound of coal to mine one pound, no steam engine could then be Promethean for that coal.[18]

Thus, coal that is available is not always accessible. Accessibility depends on the extant technology, and a Promethean technology is one that increases accessible resources to a degree beyond those used up in creating and maintaining the technology.

Georgesçu frequently criticized the "self-styled energeticists" for their overly optimistic assessments of new technological recipes, such as solar power.[19] He described solar power technology as "parasitic," that is, as requiring greater "inputs" than it offered in "outputs." Although the reader may well feel that he or she intuitively understands this notion, great caution, for the reasons explained in detail in Chapter 5, must be used in applying Georgesçu's criteria. In a paper in 1977, Georgesçu does offer very precise definitions of viable (Promethean) and parasitic technologies in terms of a primal form, input–output table for a "technology based on the direct use of solar energy."[20] He derives a familiar type of inequality condition for the technological coefficients of a Leontief (input–output) representation of an economy with three sectors ("activities" or "processes"). First, a process P_1 collects solar energy using collectors and other capital. A second sector P_2 produces collectors using solar energy and capital, while the third activity produces capital equipment using solar energy (and is completely integrated to include mining, and so on). Fund elements such as Ricardian land are omitted for simplicity. Georgesçu states (n. 20 that, "For that technology to be viable, we must have $x_{11} - x_{12} - x_{13} > 0, - x_{31} - x_{32} + x_{33} > 0$," where $x_{21} = x_{22}$ and x_{ij} is the flow of solar energy $i = (1)$, collectors $i = (2)$, or capital $i = (3)$, in sector P_1 (solar collection), P_2 (collector fabrication), or P_3 (capital fabrication). Thus, the requirement that $x_{11} - x_{12} - x_{13} > 0$ demands that solar energy collected by sector 1, x_{11}, must exceed that used in sectors 2 and 3 to fabricate equipment needed by sector 1. This definition is as precise as Georgesçu offers for a "viable technology," that is, a technology supported by a "Promethean recipe."

Note that even a viable technology does absolutely nothing to create resources. A "viable technology" is merely one that can harvest more resources than are required in its construction and operation. Economic activity is still totally dependent on the environment.

Georgesçu-Roegen was an avid student of history, and his discussions of Promethean technologies were often punctuated with historical references.[21] He felt that history provided substantial support for his thesis. A particularly favored example was the "wood crisis" that befell England and much of western Europe in the Middle Ages, when the requirements for firewood led to widespread deforestation and social upheaval.[22] Georgesçu clearly believed that our current economy, based on "heat engines" (that is, devices that convert heat differentials to work) may well meet a similar fate. The depletion of oil reserves would play much the same role as deforestation. Georgesçu felt that such an exhaustion, in the absence of a new Prometheus, would lead to disaster: missiles will fly over the last barrel of oil, and "It cannot be ruled out that some of the last people should die in penthouses, the others in hovels. Chi vivra verra."[23]

Georgesçu-Roegen's point is reinforced when, with him, we contrast the finite terrestrial *stocks* of such substances as oil or uranium with that greatest of all energy sources, the sun. As Georgesçu was fond of pointing out, the total energy contained in all the world's coal reserves amounts to but two weeks' worth of solar radiation.[24] Yet, sunlight arrives at a constant, unalterable rate, while terrestrial stocks (such as oil deposits) can, in principle, be used up at any rate selected. Man is thus induced to exploit the latter resource since he can do nothing to appropriate the former. In Georgesçu's memorable phrase, this fact has a terrible consequence: "This is why in bioeconomics we must emphasize that every Cadillac . . . means fewer plowshares for some future generations, and implicitly, fewer future human beings, too."[25]

7.4   BIOECONOMICS, PUBLIC POLICY, AND POLITICS

If one accepts Georgesçu-Roegen's characterization of economic activity, then truly Mankind can be said to face a dilemma, or "entropic predicament." Economic activity requires available energy and propitiously situated "low-entropy" materials. Production transforms these into bound energy and "high-entropy," often toxic, waste. This process is irreversible, and it results in a steady decline in the ability of the Earth to support future generations.

Despite the grim tone of this analysis, economic activity is not *only* a physical process of manipulating materials with available energy. Georgesçu-Roegen strongly believed that the primary outcome of economic activity was the "enjoyment of life" experienced by every living being. Anything that contributes to such a "flux of enjoyment" has value.[26]

Georgesçu contrasts his views on the notion of economic "value" with those of the Marxists and other classicals through an extended analysis of a "general equation of value."[27] The symbolic template of this analysis can be given as $e = CE + LE - WD$, where e is an individual's "daily life enjoyment," CE his consumption enjoyment, LE leisure enjoyment, and WD work drudgery. Of course, the addition sign and similar mathematical symbols must be interpreted only as "convenient signs for summarizing the imponderable elements that enter . . . into the activity represented on the left of the equality sign."[28]

Georgesçu argues that everything which contributes to the enjoyment of life, directly or indirectly, has an economic value. This view is a departure from some neoclassical usages since prices, for example, depend on consumption excludability criteria that cannot be applied to such basic essentials as sunlight. Georgesçu argues that Ricardo, Marx, and other classical economists, in their conceptual presentations of value theory, implicitly or (in the case of Marx) explicitly ignored important components of value.[29] Ricardo, for example, sometimes seemed to argue that leisure had no economic value. Marx's scheme appeared to Georgesçu to exhibit a similar flaw, in addition to a host of other flaws identified by other writers. He reserves his harshest criticism for the neo-classicals, however, and roundly attacks most neoclassical analyses for neglecting the value of leisure, and further failing to take account of the disutility of work, a separate issue. All of this can be taken as a closely reasoned plea to incorporate the value of leisure in the national accounts. However, the issue is more complex than might first appear because overpopulated agrarian societies exhibit substantial *unwanted* leisure, an issue we addressed in chapter 4.

Thus, although Georgesçu stresses the physical aspects of economic activity, he accepts the proposition that human welfare is the sole goal.[30]

Given the entropic nature of economic activity, what, then, should we do? Which policies or political principles are required in order to "love thy species as thyself," as Georgesçu sometimes recommended? Because Georgesçu-Roegen was first and foremost an academic economist and scholar, writing for an academic audience, he believed that economists, and all academics, inherited large and undischarged obligations representing the debts owed by any elite towards the working classes whose toil afforded scholars the luxury of the intellectual life. This bargain is imposed by morality, and Georgesçu felt it so keenly that he damaged his career, and nearly lost his life, by returning to Romania in the 1930s. He continued to fulfill this bargain as best he could through his work with Dai Dong, his writing on development economics, agrarian reform, and population, and in public advocacy of his positions. Thus, Georgesçu was both willing and capable of presenting policy recommendations, and did so in numerous forums. We now review several of his social recommendations.

A fundamental proposition of bioeconomics, as formulated by Georgesçu, is that current generations overuse resources. This occurs because future generations cannot contract in current markets. Unless someone acts as an agent for posterity, posterity receives "too few" natural resources.

We do not address here the validity of this argument. Rather, we note only that, if one accepts it, then interventions to slow the exploitation of resources may be warranted. Georgesçu proposed several ways to achieve this end. First, he suggests that we combine population reductions with an ascetic lifestyle, eliminating fads, "disposable" goods, and idle luxuries. Georgesçu stated: "Mankind should gradually lower its population to a level that would be adequately fed only by organic agriculture," and later: "we must cure ourselves of the morbid craving for extravagant gadgetry, splendidly illustrated by such a contradictory item as a golf cart . . . we must also get rid of fashion, of that 'disease of the human mind' . . ."[31]

Georgesçu-Roegen's other suggestions are of a similar character. He variously proposes that we (a) conserve resources; (b) abolish war and, in particular, armaments; (c) abolish visas and allow for the free flow of persons across national borders; (d) "globalize" the ownership of some resources; (e) move rapidly to facilitate adequate improvements in the living standards of poor countries.[32]

Georgesçu-Roegen's analyses of population are especially interesting and noteworthy. Although this was touched on in Chapter 3, it is appropriate here to reexamine his insight. Georgesçu was interested not just (or primarily) in "how many?", but in "for how long?". Although the Earth may be capable of supporting 50 billion people for a year, it may be capable only of supporting 1 billion people for 5 million years. In Georgesçu-Roegen's view, the greater the population, the faster the "entropic degradation" of the world will occur. Because our solar "fund," the sun, can be expected to provide us with energy for billions of years, the long-run interests of the human race would appear to require a modest population maintained for a very long time. Only in this way do we avoid forgoing the benefits of the solar radiation of tomorrow. As Georgesçu said: "to have a maximum population at all times is definitely not in the interest of our species. The population problem . . . concerns the maximum of the life quantity that can be supported by man's natural dowry."[33]

As one might expect of a man who witnessed and survived the Second World War, fascism, and Stalinist communism, Georgesçu was not eager to propose the government controls that could be necessary in order to realize his policy suggestions. One might then plausibly argue that his proposals, such as the abolition of war, amount to no more than a utopian dream. Even the far more realistic suggestion that we conserve resources (that is, utilize them at rates below those induced by the price system alone) will presumably require some type of intervention.

In reply to these criticisms, there are two responses evident in Georgesçu's writings. First, he occasionally proposed direct government regulation of resource depletion rates.[34] Second, and more importantly, the entire purpose of Georgesçu's environmental advocacy was to *change people's minds*. For example, his constant criticisms of mechanical analogies in economic theory, and of the neoclassical treatments of environmental inputs, were made precisely because he believed such patterns of thought were a substantial barrier to accurate thinking about the future of Mankind. Georgesçu's criticisms of statistical thermodynamics were not offered to further the status of Georgesçu-Roegen, but because an unwarranted belief in "entropy bootlegging" made it possible for brilliant economists, such as Solow, to suggest that we (humans) could "get along without natural resources."[35] As Georgesçu sometimes described it, Man clings tenaciously to the proposition that, "Come what may, we will find a way."

Georgesçu-Roegen believed that history is an evolutionary process. As with any evolutionary process, extrapolation, a type of "linear thinking," cannot be validly used to forecast the future. Only mechanical processes are liable to such extrapolation. Thus, the fact that humans are, in many ways, living more comfortably now than in the past, and that critical shortages of minerals are not commonly observed, cannot by themselves be used to credibly claim that the future, even the near future, will be equally pleasant. Georgesçu strongly argued that the last 100 years were very atypical, and termed our recent history a period of "mineralogical bonanza."[36] By adopting, if only unconsciously, a mechanical representation of economic activity, we fall into the trap of believing the future will be like the past. If Nicholas Georgesçu-Roegen is correct, this is a fatal delusion.

7.5   REFLECTIONS ON BIOECONOMICS

The description of bioeconomics given in the preceding sections, while containing what we regard as the most important elements, is incomplete. In particular, two important issues remain unexamined. The first, on which only a little space will be expended, concerns economic *models* that incorporate bioeconomic elements. Georgesçu-Roegen made only limited efforts in this direction, but some recent work, such as that of Faber *et al.* (1987/1995), may point the way towards *implementing* bioeconomics in applied work.

Second, missing from much (though not all) of Georgesçu-Roegen's bioeconomic writing is any clear treatment of pollution, toxic waste, biodiversity – in short, the *outputs* of economic activity.[37] Rather, one encounters in Georgesçu's writings a preoccupation with the *inputs*. Obviously, both the availability of inputs and the toxicity of outputs are critical to human welfare.

Georgesçu-Roegen's emphasis on inputs probably arose from several sources. First, the publication of much of his most important work on bioeconomic themes, beginning with *The Entropy Law and the Economic Process* in 1971 and continuing through "Energy analysis and economic valuation" in 1979, coincided with the oil shortages of the 1970s, although of course his work on entropy goes back in print to the mid-1960s. Georgesçu-Roegen frequently mentioned the oil crisis in his writings of this period.[38]

It is also true that Georgesçu was far less informed on biological topics than on thermodynamics, despite his frequent, glowing references to Alfred Marshall's assertion that biology, not physics, was the "Mecca of the economist." References to biology in Georgesçu's writings are frequently as dated as those to physics, yet clearly biology and ecology have changed more than classical thermodynamics between 1945 and the late 1970s. Biological science is critical for understanding many aspects of the problems of pollution and species extinction. It can also be noted that general scientific concern for some of these issues, such as biodiversity, is relatively recent.

Finally, our earlier discussion of Georgesçu-Roegen's methodological position highlighted the fact that one may interpret his message, simply put, as "the economic process is one in which stocks of some things are used up and stocks of other things accumulate." The stock that is "used up" refers, of course, to "low-entropy," useful materials, such as minerals, which are dispersed to unusable concentrations over time. The stocks that accumulate include waste products, mine tailings, thermal pollution, and so on. These latter stocks are generally harmful, both to individuals through toxic effects and to species and ecosystems through the loss of habitat.

There is a relatively important difference between the accumulating and depleting stocks. While concentrating and refining minerals, such as aluminum, involves minimal energy requirements imposed by the Entropy Law, and while the dispersion of materials is analogous to a particular notion of entropy, the analysis of toxic waste is not readily facilitated by reference to the Entropy Law. Many substances that constitute toxic wastes are "low-entropy" materials, which are costly to create. A plausible consequence of this is that Georgesçu focused his efforts on the input side of the ledger because that side was most relevant to his recognition of the importance of entropy for economic life.

Bioeconomics remains a work in progress. Georgesçu's effort in "Energy analysis and economic valuation" (1979) is perhaps his foremost attempt to operationalize at least some of his ideas. In this and a few similar analyses, he utilizes his flow-fund model to represent a "bioeconomy," in which resources provided by the environment, and recycling, are explicitly included. A more sophisticated and satisfactory bioeconomy is modeled by Faber *et al.* (1995). Despite the sophistication of these and similar efforts, no one can claim that the last words on bioeconomics have been written. Likewise, no one can deny that Nicholas Georgesçu-Roegen penned the first.

NOTES

1. Georgesçu credits Jiri Zeman for coining this word. See "Energy and economic myths," 1975: 369, n. 50.
2. Georgesçu-Roegen in Szenberg (1992: 146).
3. We examine this idea in detail in Chapter 3.
4. Daly (1992) provides considerable elaboration on Daly's views.
5. In particular, the "Fourth Law" is claimed to rule out perfect recycling.
6. In Georgesçu's words, "such a technology if it is to be viable must be capable of reproducing itself after being set up by the technology now in use" ("Technological assessment: the case of the direct use of solar energy", 1978: 18).
7. Daly (1992) provides a frank discussion of these ideas.
8. *The Entropy Law and the Economic Process* (1971: ch. 10, s. 4) provides a detailed discussion.
9. Ibid.: 310–12.
10. Virtually all of Georgesçu-Roegen's discussions of bioeconomics after 1972 include something on this theme. See "Energy analysis and economic valuation" (1979) for a representative treatment.
11. "[L]ike Marx, I believe that the social conflict is not a mere creation of man without any root in material human conditions. But unlike Marx, I consider that, precisely because the conflict has such a basis, it can be eliminated neither by man's decision to do so nor by the social evolution of mankind" (*The Entropy Law*, 1971: 306).
12. See also ibid.: appendix G. Georgesçu's knowledge of biology was not very complete.
13. "Energy and economic myths" (1975: 328).
14. Scholars, unfortunately, do not always share in this bounty.
15. See Rees, "The theory of principal and agent," in Hey and Lambert (1989) for a good survey.
16. We treat this in Chapter 4.
17. Georgesçu-Roegen in Szenberg (1992: 150).
18. Ibid.: 151.
19. Georgesçu wrote: "The current view that only energy matters for mankind's specific mode of existence is not completely identical with that school's position, but at bottom the two are sufficiently similar to justify the term 'energetics' for its label" ("Energy analysis and economic valuation", 1979: 1024). Georgesçu refers here to the scientific work of Wilhelm Ostwald and some others.
20. "Technology assessment: the case of the direct use of solar energy" (1978). A similar model, though far more complete, is offered in "Energy analysis and economic valuation" (1979).
21. This interest in history stemmed, in part, from Schumpeter's tutelage. Georgesçu was not averse to using historical facts to bolster his arguments, but he was not willing to use such data to attempt to derive *analytic* laws.
22. Braudel (1979: 366–7) gives an interesting account of these early "energy crises."
23. Georgesçu-Roegen in Szenberg (1992: 152).
24. See "Energy and economic myths" (1975: s. IX) for this estimate.
25. Ibid.: 370.
26. While the "flux" varies in intensity, no stock of it accumulates or decumulates, again illustrating the importance of abandoning the neoclassical meaning of such terms.
27. *The Entropy Law*, ch. 10, s. 2.
28. Ibid.: 285.
29. The "cornerstone of Marx's doctrine," according to Georgesçu, is the notion that "nothing can have value if it is not due to human labor" (ibid.: 289).
30. Georgesçu-Roegen, unlike some modern ecologists, never refers either to "rights" held by animals or to "intrinsic" values of ecosystems.

31. "Energy and economic myths", 378.
32. A good review of most of Georgesçu's proposals occurs in ibid.
33. *The Entropy Law and the Economic Process*, (1971: 20).
34. Ibid.: 378.
35. Solow (1974: 11) remarks: "if the elasticity of substitution between exhaustible resources and other inputs is unity or bigger . . . then a constant population can maintain a positive constant level of consumption per head forever."
36. This term is taken to mean that all current production comes from mines of the highest (or higher) grades.
37. However, in "Energy and economic myths" Georgesçu notes that, "Because pollution is a surface phenomenon . . . we may rest assured that it will receive more official attention than its inseparable companion, resource depletion" (p. 377).
38. Georgesçu did not, however, point to the "energy crisis" as somehow implying that *energy alone* would solve our problems.

8.   Conclusion

In philosophy, Boltzmann advocated the complementarity of contradictory
hypotheses, according to which contradictory theories of the natural world
may yet all be correct. Such theories should be regarded as complementary
rather than antagonistic. It is remarkable that Boltzmann's thinking here
adumbrated so clearly Niels Bohr's later "principle of complementarity"
based on quantum theory in which, to properly describe the world, one must
be able to combine mutually exclusive concepts with varying degrees of
probability.

Coveney and Highfield, *The Arrow of Time*

Dialectics rather than arithmomorphism is my creed.

Nicholas Georgesçu-Roegen

8.1   SUMMARY

Nicholas Georgesçu-Roegen, as Screpanti and Zamagni (1993) observe, is
difficult to classify.[1] This observation is both appropriate and ironic. Classification, as science ordinarily uses this term, is of a binary nature. Nature, with
a capital "N," is not. The recognition of this fact informed every aspect of
Nicholas Georgesçu-Roegen's professional life. No mysticism was invoked by
Georgesçu, nor is mysticism required to understand him. When he urged us to
recognize that mechanical analogies were incapable of representing most of
reality, he was not being a shill for pseudo-science. Rather, he was reporting
and – in an unfortunately obscure way – explaining the "true facts" of the real
world in which we live. Georgesçu-Roegen's invocation of "object lessons
from physics" had as a primary goal convincing economists that the mechanistic
dogma was not preeminent in science. Such a recognition was needed, he felt,
to transform economics into a field useful for discussing the destiny of
humankind.

Georgesçu had a very long career: his first publications (in mathematics)
appeared in the 1920s, and his final work was published in 1993, one year
before his death. Georgesçu wrote almost 200 works, including articles, books,
reports, and invited lectures. As one would expect, such an opus is bound to
contain a lot that is right, some that is wrong, and some that must fall into the
"undecidable" category. If our study of the work of Nicholas Georgesçu-Roegen
has produced a surprise for us, it is undoubtedly in the paucity of things that are

wrong. Debatable conjectures there are in abundance (the Fourth Law being only the best known), but inaccuracies are remarkably few.

Georgesçu's extensive intellectual legacy in print is, to an incredible degree, internally consistent, which seems even more remarkable when one recognizes the great period of time over which it was produced. If one sought for a single word to encapsulate the unity in Georgesçu's vision, that word would be "Evolution." Evolution – true change – was, for Georgesçu-Roegen, the starting and ending point for the vast majority of his work.

The role of the idea of evolution in Georgesçu's thinking cannot be overstated. His methodology and dialectics take the evolutionary character of phenomena as a brute axiom. Evolution is not a mechanical process, and it cannot be given arithmomorphic representations. Analytic relations between real numbers are a language inadequate to express critical aspects of natural and economic processes. The arithmetic continuum is merely "beads on a string," without the string. Discrete distinctness, the defining property of real numbers, cannot be fruitfully applied to many critical concepts in economics and science.

It is impossible to discuss evolution, or propose an evolutionary law, without invoking the notion of time. One could fairly say that Georgesçu-Roegen's signature preoccupation was time, rather than evolution, since these concepts were, to Georgesçu, inseparable. Evolutionary laws, such as the Second Law of Thermodynamics, give us "time's arrow." The processes of the natural world are *not* time symmetric. Except in an exceptional minority of cases, analytic, arithmomorphic models cannot correctly represent any actually occurring physical process.

The twin concepts of evolution and time informed almost all of Georgesçu-Roegen's work. His early and important analyses in consumer theory ask us to suppose that consumer choices alter preferences. Wants are not banished from utility theory merely because "want" is a dialectical concept. Wants exhibit a hierarchical structure, and the Principle of Substitution is not assumed. Expectations are not merely numbers, and our attempt to represent them as such is misleading. In every case, arithmomorphic representations of human behavior and the human mind are challenged, and alternatives are offered in their places.

Production theory, particularly the flow-fund representations Georgesçu invented, brings the role of time and evolution in Georgesçu's work to center stage. Time and *order* are emphasized. The cavalier treatment of time in neo-classical models long obscured the important distinction between stocks and funds. Time appears as an explicit right-hand-side variable in Georgesçu's production models. Production is a process, requires a duration even to define, and is subject to limitations on substitution among factors imposed by the Entropy Law.

Production in agriculture differs fundamentally from the "factory system" in that the former process cannot be started at arbitrary times. This simple fact, Georgesçu felt, had profound implications for economic development and the sociology of urban–rural relations.

Georgesçu-Roegen applied his ideas in production theory to examine the historical development of Mankind. Georgesçu termed a technology "Promethean" if it could serve as the basis of an ongoing social system. Such a technology must be "viable," that is, it must be capable of producing more "resources" than its fabrication and maintenance consumes, with a surplus left over. Promethean technologies cited by Georgesçu are animal husbandry, fire, and heat engines.

The notion of a Promethean technology is important in Georgesçu's bioeconomic program. Further, he felt that the necessity of Promethean technologies imposed an evolutionary character on human economic development. Any Promethean technology requires, as all economic activities do, a steady diet of natural resources: fire requires wood, steam engines require coal, and so on. Human propensities being what they are, humankind soon finds itself faced with a "crisis" arising from the growing scarcity of accessible stocks of the needed resources. These crises trigger social changes and, perhaps, a "new Prometheus."

Yet history is an evolutionary process and, like all such processes, one cannot extrapolate to determine the future course of events. Discovery of a new Promethean technology involves luck and accident: the cost of being unlucky is not pleasant to contemplate.

Although Georgesçu sometimes stated that no "evolutionary laws of society" had been discovered, he clearly believed that evolutionary laws, especially the Entropy Law, substantially and materially affected social evolution. For example, the Entropy Law implies that there is a strict minimum energy requirement to transform materials and, although this energy requirement can, and often is, satisfied by commodities that are not traded in energy markets, this constraint is absolute and will never be overturned by technological advances. As ores are refined, for example, and lower and lower grades of ore are exploited, these energy requirements will increase. Notions of unlimited substitution between inputs, a doctrine adopted by many technological optimists, cannot be literally true.

Georgesçu-Roegen also believed that "matter matters, too." In particular, he believed and frequently stated that matter in useful, ordered structures underwent irrevocable dissipation and became widely distributed in the environment. Although he claimed in some cases that the Entropy Law applied to ordered structures, he later came to realize that this was not correct in the sense he intended. While only materials have an entropy, the Entropy Law, which does establish that free energy degrades into bound energy, does not imply anything similar for "available" matter. Hence, Georgesçu proposed the "Fourth Law of Thermodynamics": a closed system, receiving energy (but not matter) from outside, cannot perform work at a constant rate forever. This law, if it is a law, can be paraphrased as "Perfect recycling is impossible."

Georgesçu-Roegen's observation on the dissipation of matter is, as a practical issue, ecologically relevant even if the Fourth Law is false. Recycling will presumably always be imperfect in practice, even if perfection is not ruled out in theory. Given this, propitiously situated material structures will decline over time, leaving future generations with less than the current generation enjoys.

The depletion of resources is important for politics and ethics: how, and to what extent, should the behavior of current generations be altered to preserve resources for the future inhabitants of our world? Contained within this question are several separate and important issues examined by Georgesçu. First, is there a moral obligation extending from the present to the future? Second, is it likely that the price system, left to itself, will use up resources "too fast"? Finally, what policies might be implemented to slow resource depletion should such a course be warranted?

Georgesçu-Roegen repeatedly argued that: (a) we have a moral obligation to future generations; (b) because there is no substitute for natural resources, our obligations must be met by limiting our use of our terrestrial dowry; and (c) the price system, left to itself, will result in "excessive" exploitation of natural resources.

A careful reading of "Energy and economic myths" (1975) indicates that Georgesçu's analysis of this issue relies on several "facts." First, some minimum amount of resources is always required by each generation. Second, resources are consumed sequentially from "best" to "worst" quality (concentration). Finally, demands for resources by each generation are "parochial" in the sense that the bequest motive, overlapping generations, and other similar mechanisms are insufficient to cause current period demands to fully reflect the demands of future generations. In other words, the simple physical impossibility of future generations trading in current resource markets is not completely mitigated by pure market incentives to conserve resources.

Georgesçu believed that minimal necessary quantities of resources should be used up by each generation, and that use above this level represented an immoral "dictatorship of the present over the future."[2]

It is important to recognize that the whole issue of intergenerational equity is a testament to the importance of Georgesçu's work. Natural resources are special and are unlike other inputs in their irreplaceable quality. Recycling is imperfect and may well always be so. Under these conditions important intergenerational issues arise.

The culmination of Georgesçu-Roegen's concerns with resource depletion and intergenerational equity was the "bioeconomic program," Georgesçu's ambitious attempt to remake economics into a discipline useful for discussing the long-term prospects of humanity. Bioeconomics, as it came to be called, combined sociological theorizing with thermodynamics, the price theory of exhaustible resources, and politics.

Georgesçu-Roegen, whatever his faults as an analyst (and they were few), created a powerful indictment of the prevailing neoclassical paradigm in economics. Of course, this indictment need not debilitate mainstream work in economics; with Stiglitz, we may remark: "I am not concerned with long-run problems arising from the laws of thermodynamics."[3] Likewise with Stiglitz, we may dismiss a part of Georgesçu's analysis of the intertemporal allocation of natural resources with the statement: "it is sometimes argued that because our children and grandchildren are not present today, the market will systematically underrate their importance; . . .This overlooks the fact that the investor can sell the asset . . . to another investor to obtain his return."[4] Certainly, we too may dismiss long-run problems, and most economists do so. Georgesçu-Roegen could not, and did not.

8.2   GEORGESÇU-ROEGEN: RIGHT AND WRONG

We turn now to a brief yet hazardous review of the more important conclusions of this work. The term "hazardous" is especially appropriate in this context, both because many of Georgesçu-Roegen's propositions are quite difficult, and because prevailing opinion, whether concerning matters economic or physical, can well be wrong.[5] Nevertheless, it is an important goal of this monograph to stimulate discussion and study of Georgesçu-Roegen, and that end is promoted by an unadorned review of this material. Thus, we list below several important propositions by Georgesçu-Roegen, and follow with our brief evaluation of their validity.

(A)   "The entropy law is relevant to the economics of resource scarcity"

This is correct. As illustrated in the three-chambered hourglass analogy of Chapter 6, the Second Law imposes an absolute entropy increase requirement on the economic process of transforming any material. Accompanying each entropy increase is a required degradation of energy. Energy may not be recycled, and the maintenance of a material in a state away from equilibrium requires energy. No technological advance will ever overturn these constraints. However, these requirements can be satisfied by use of materials, such as chemical agents, that are unconnected with the energy market.

(B)   "A closed system that receives inputs of energy but not matter at a constant rate cannot perform work at a constant rate forever; complete recycling is impossible"

The status of the Fourth Law is uncertain. Many biologists and ecologists seem to doubt the validity of the law. Investigations of the validity of the law are problematic.

(C)   "The H-theorem is false, and Newtonian mechanics cannot provide an arrow of time"

Georgesçu-Roegen is correct about this: the H-theorem is wrong, and time's flow is more than the greatest probability manifesting itself. The work of Ilya Prigogine and the so-called "Brussels School" in thermodynamics has introduced new concepts into this investigation, as has the work of cosmologists, and the source of time's arrow remains under debate.[6]

(D)   "Statistical thermodynamics lacks experimental support, and is incorrect"

Georgesçu-Roegen is here incorrect. As explained in Chapters 5 and 6, there is now a substantial body of experimental evidence that provides support for the statistical approach to thermodynamics. Further, statistical thermodynamics can explain the magnitudes of the standard entropies of some substances. This facility is never offered by classical thermodynamics.

His blanket rejection of statistical thermodynamics was a serious error. Statistical thermodynamics is not identical with the H-theorem, nor does it depend on it.

(E)   "The price system results in overuse of natural resources by current generations"

The issue of intergenerational equity is, first and foremost, an issue of equity. Under all the Arrow–Debreu neoclassical assumptions, competitive allocation of resources will be Pareto optimal. Yet Pareto optimality itself is quite weak and, if errors in future forecasts occur, there may be no Pareto moves from a current allocation towards an equitable one.[7] Georgesçu sometimes seemed to argue that only equal sharing of resources across time was equitable.[8] Such a view is probably not reasonable.[9]

The problem of intergenerational equity remains a fundamental one for economics and philosophy. The idea that society should not have a discount rate, though all citizens do, is completely consistent with Georgesçu's point that properties of a whole cannot merely reflect amplification or aggregation of the properties of parts. As an empirical matter, the claim that current generations use resources up "quickly," given a horizon of, say, a millennium or two, is hard to dispute.

8.3   GEORGESÇU-ROEGEN IN RETROSPECT

Finding the "right place" for Nicholas Georgesçu-Roegen will be an ongoing effort. Progress in this task will require a far greater dissemination of his work than has yet occurred. Until economists of the mainstream become effectively familiar with Georgesçu-Roegen, an event that will require the painful study of thermodynamics, it is probable that he will remain a heretic whose work, when it is presented at all, will be placed at the "end of the book."

In all fairness, one can plausibly argue that Georgesçu belongs at the "end of the book." After all, it is not difficult to claim that neoclassical economics, with its arithmomorphic models and narrow focus, has been quite effective in analyzing many important problems. The combination of maximization principles, constraint sets, analytic representations of utility, production relationships, and expectations is powerful. The importance of correctly analyzing "the best means to a *given* end" is undeniable. Perhaps the future will take care of itself, as it has in the past.

Yet if scholars such as Kenneth Boulding, John Ise, Alfred Lotka, and Nicholas Georgesçu-Roegen make us doubt, or make us a little uncomfortable, we take that as a very good thing. Georgesçu-Roegen, in particular, leaves his mark on the reader, and is capable of producing a lingering dissatisfaction with economics as usual.

The long-run survival of the human race is an immensely important topic. The uncertainties involved in this problem are profound, as Nicholas Georgesçu-Roegen would be the first to admit. But it is not correct to say that we know nothing about the character of the technology of the future. The energy requirements for transforming materials imposed by the Entropy Law, though not simple arithmomorphic constraints, are real constraints nevertheless. The fact that the solar system's entropy is irrevocably rising stamps a unidirectional character on economic development which the standard "circular flow" model of the economy completely misses. This knowledge must inform our discussions of the future. In fact, these constraints are a far more valuable insight than any one can obtain merely by extrapolating from the recent past. It is undeniable that history is an evolutionary process and, as such, cannot be forecast by extrapolation. Insights on which one can reliably depend are surely valuable.

We conclude with a plea. Reading Nicholas Georgesçu-Roegen is a demanding and often frustrating task. It is far easier to rely on accounts, such as this book, than on original source material. This tendency should not be gratified. We urge the reader to go to the original works. The payoff justifies the effort.

NOTES

1. Screpanti and Zamagni (1993: 417).
2. "Energy and economic myths" (1975: 375).
3. Stiglitz in Smith (1979: 37).
4. Ibid.: 52.
5. In particular, the history of Boltzman's H-theorem should serve as a warning.
6. A fascinating popular account of this issue, and Prigogine's contributions, is in Coveney and Highfield (1990).
7. See Howarth and Norgaard in Bromley (1995) for an exposition.
8. Alternately, he sometimes argued for a criterion he termed the "principle of minimizing regrets." See "Energy and economic myths" (1975).
9. In particular, such a plan would appear inefficient in the presence of technological progress.

Selected bibliography of Georgesçu-Roegen

Note: This listing omits works in Romanian from the war period (1939–45), some unpublished works, reprinted works except in important cases, and non-academic work such as letters to newspapers. Some of these omitted materials are listed in *Evolution, Welfare, and Time in Economics* (Tang *et al.*, 1976).

"Sur un problème de calcul des probabilités avec application à la recherche des périodes inconnues d'un phénomène cyclique." *Comptes Rendus de l'Académie des Science* **191** (July 7, 1930): 15–17.

"Le problème de la recherche des composantes cycliques d'un phénomène." Dissertation. *Journal de la Société de Statistique de Paris* (October 1930): 5–52.

"Further contributions to the sampling problem." *Biometrika* (May 1932): 65–107.

"Sur la meilleure valeur a posteriori d'une variable aléatoire." *Bulletin de la Société Roumaine des Sciences* **33–34** (1932): 32–8.

"Tehnica numerelor indice pentru nivelul general al preturilor." *Buletinal Statistic al României* **4** (1932).

"Note on a proposition of Pareto." *Quarterly Journal of Economics* (August 1935): 706–14.

"Fixed coefficients of production and the marginal productivity." *Review of Economic Studies* (October 1935): 40–9.

"Marginal utility of money and elasticities of demand." *Quarterly Journal of Economics* (May 1936): 533–9.

"The pure theory of consumer's behavior." *Quarterly Journal of Economics* (August 1936): 545–93.

"L'influence de l'age maternel, du rang de naissance, et de l'ordre des naissances sur la mortinalité" (with R. Turpin and A. Caratzali). *Premier congrès Latin d'Eugénique* (1937): 271–7.

"Further contributions to the scatter analysis." *Proceedings of the International Statistical Conferences* **5** (1947): 39–43.

"A supra Problemei 5527." *Gazeta Matematicã* **52**, Commemorative issue (October 1947): 126–33.

"Further contributions to the scatter analysis," *Econometrica* **16** (1948): 40–3.

"The theory of choice and the constancy of economic laws." *Quarterly Journal of Economics* **64** (February 1950): 125–38.

"Leontief's system in the light of recent results." *Review of Economics and Statistics* (August 1950): 214–22.

"Some properties of a generalized Leontief Model." In Koopmans, T., ed., *Activity Analysis of Production and Allocation*. New York: Wiley, 1951: 165–73.

Activity Analysis of Production and Allocation (coeditor with T.C. Koopmans *et al.*). New York: John Wiley, 1951.

Review of W.W. Leontief, *The Structure of American Economy*. *Econometrica*, **19** (1951): 351–3.

"A diagrammatic analysis of complementarity." *Southern Economic Journal* (July 1952): 1–20.

"Toward partial redirection of econometrics." *Review of Economics and Statistics* (August 1952): 206–11.

"Note on Holley's dynamic model." *Econometrica* (July 1953): 457–9.

"Multi-part economic models: discussion." *Econometrica* (July 1953): 469–70.

"Note on the economic equilibrium for nonlinear models." *Econometrica* (January 1954): 54–7.

"The end of the probability syllogism?" *Philosophical Studies* (February 1954): 31–2.

"Choice and revealed preference." *Southern Economic Journal* (October 1954): 119–30.

"Choice, expectations and measurability." *Quarterly Journal of Economics* (November 1954): 503–34.

"Limitationality, limitativeness, and economic equilibrium." *Proceedings of the Second Symposium in Linear Programming*, vol. I. Washington, D.C.: 1955, pp. 295–330.

Review of Maurice Allais, *Traité d'économie pure*. *American Economic Review* **46** (1956): 163–6.

"Economic activity analysis" (review article). *Southern Economics Journal* **22** (1956): 468–75.

"The nature of expectation and uncertainty." In *Expectations, Uncertainty, and Business Behavior*, ed. Mary Jean Bowman, New York: Social Science Research Council, 1958, pp. 11–29.

"Threshold in choice and the theory of demand." *Econometrica* **26** (January 1958): 157–68.

"On the extrema of some statistical coefficients." *Metron* **19** (July 1959): 1–10.

"Economic theory and agrarian economics." *Oxford Economic Papers*, N.S., **12** (February 1960): 1–40.

"Mathematical proofs of the breakdown of capitalism." *Econometrica* **28** (April 1960): 225–43.

Review of E. Levy, *Analyse structurale et méthodologie économique*. *American Economic Review* **52** (1962): 1123–4.

"Some thoughts on growth models: a reply." *Econometrica* **31** (January–April 1963): 230–6, 239.

"Measure, quality, and optimum scale." In *Essays on Econometrics and Planning Presented to Professor P.C. Mahalanobis*. Oxford: Pergamon Press, 1964, pp. 231–56.

Review of Michio Morishima, *Equilibrium, Stability and Growth: A Multisectoral Analysis*. *American Economic Review* **55** (1965): 194–8.

Analytical Economics: Issues and Problems. Cambridge, MA: Harvard University Press, 1966.

"Further thoughts on Corrado Gini's *Delusioni dell'econometria*." *Metron* **25** (1966): 265–79.

"Chamberlin's new economics and the unit of production." In *Monopolistic Competition: Studies in Impact. Essays in honor of Edward H. Chamberlin*, ed. R.E. Kuenne. New York: John Wiley, 1967, pp. 31–62.

"An epistemological analysis of statistics as the science of rational guessing." *Acta Logica* **10** (1967): 61–91.

"O estrangulamento: inflação estrutural e o crecimento econômico." *Revista Brasileira de Economia* **22** (March 1968): 5–14.

"Utility." In *International Encyclopedia of Social Sciences*, vol. 16. New York: Macmillan and Free Press, 1968, pp. 236–67.

"Revisiting Marshall's constancy of marginal utility of money." *Southern Economic Journal* **35** (1968): 176–81.

"Process in farming versus process in manufacturing: a problem of balanced development." In Ugo Papi and Charles Nunn, eds, *Economic Problems of Agriculture in Industrial Societies*. Proceedings of a conference held by the International Economic Association, Rome, 1969.) London: Macmillan, 1969, pp. 497–528.

"A critique of statistical principles in relation to social phenomena." *Sociological Abstracts* **17**, no. 5, suppl. 6 (August 1969): 9.

"Relations between binary and multiple choices: some comments and further results." *Econometrica* **37**, no. 4 (October 1969): 726–8.

"The institutional aspects of peasant communities: an analytical view." In *Subsistence Agriculture and Economic Development*, ed. Clifton R. Wharton, Jr. Chicago: Aldine, 1969, pp. 61–99.

"Structural inflation-lock and balanced growth." *Economies et Sociétés, Cahiers de l'Institut de Science Economique Appliquée* **4**, no. 3 (March 1970): 557–605.

"Technology assessment: the case of the direct use of solar energy", *Atlantic Economic Journal* **6** (4), 1970: 18.

"The economics of production" Richard T. Ely lecture. *American Economic Review, Papers and Proceedings* **60**, no. 2 (May 1970): 1–9.

The Entropy Law and the Economic Process. Cambridge, Mass.: Harvard University Press, 1971.

"Analysis versus dialectics in economics." In *Ensaios Econômicos, Homagem a Octávio Gouvêa de Bulhões*, ed. Micea Buescu. Rio de Janeiro: APEC, 1972, pp. 251–78.

"Process analysis and the neoclassical theory of production." *American Journal of Agricultural Economics* **54**, no. 2 (May 1972): 279–94.

Analisi economica e processo economico. Florence: Sansoni, 1973.

"Utility and value in economic thought." *Dictionary of the History of Ideas.* New York: Scribner's, 1973, vol. 4, pp. 450–8.

"The Entropy Law and the economic problem." In *Toward a Steady-state Economy*, ed. Herman E. Daly. San Francisco: W.H. Freeman, 1973, pp. 37–49.

Review of John S. Chipman, Leonid Hurwicz, Marcel K. Richter, and Hugo F. Sonnenschein, *Preferences, Utility, and Demand. Journal of Economic Literature* **11**, no. 2 (June 1973): 528–32.

"Mechanistic dogma and economics." *Methodology and Science* **7**, no. 3 (1974): 174–84.

"L'economia politica come estensione della biologia." *Note Economiche*, ed. Monte dei Paschi di Siena, 1974, no. 2, pp. 5–20.

"Dynamic models and economic growth." *Economie appliquée* **27**, no. 4 (1974): 529–63.

"Energy and economic myths." *Southern Economic Journal* **41**, no. 3 (January 1975): 347–81.

"The Entropy Law and economics." In *Entropy and Information in Science and Philosophy*, ed. Jiři Zeman. Amsterdam: Elsevier, 1975, pp. 125–42.

"Energy and economic myths." *The Ecologist* **5**, nos 5 and 7 (June and August–September 1975): 164–74, 242–55.

"Dynamic models and economic growth." *World Development* **3** (November–December 1975): 765–83.

"Vilfredo Pareto and his theory of ophelimity." In *Convegno Internazionale Vilfredo Pareto* (Rome, October 25–27, 1975). Rome: Academia Nazionale dei Lincei, 1975, pp. 223–65.

Energy and Economic Myths: Institutional and Analytical Economic Essays. Oxford: Pergamon Press, 1976.

"Economic growth and its representation by models." *Atlantic Economic Journal* **4**, no. 1 (Winter 1976): 1–8.

"Economics and mankind's ecological problem." In *U.S. Economic Growth from 1976 to 1986: Prospects, Problems, and Patterns*, vol. 7, Washington, D.C.: Joint Economic Committee, Congress of the United States, 1976.

"Is the perpetual motion of the third kind possible?" (mimeo). Paper read at the Colloquium of the ENST and University of Paris IX, November 19, 1976.

"Matter matters, too." In *Prospects for Growth: Changing Expectations for the Future*, ed. K.D. Wilson. New York: Praeger, 1977.

"What thermodynamics and biology can teach economics." *Atlantic Economic Journal*, 1 (March 1977): 13–21.

"The steady state and ecological salvation: a thermodynamic analysis." *BioScience*, April 1977.

"Bioeconomics: a new look at the nature of the economic activity." In *The Political Economy of Food and Energy*, ed. Louis Junker. Ann Arbor, Mich.: University of Michigan, 1977.

"Matter: a resource ignored by thermodynamics." In *Proceedings of the World Conference on Future Sources of Organic Raw Materials*, Toronto, July 10–13, 1978, New York: Pergamon, 1978.

"Technology assessment: the case of the direct use of solar energy." *Atlantic Economic Journal*, 1 (December 1978): 15–21.

"Comments on the papers by Daly and Stiglitz." In *Scarcity and Growth Reconsidered*, ed. V.K. Smith. Baltimore, Md: Johns Hopkins University Press, 1979, pp. 95–105.

"Myths about energy and matter," Lexington Conference on Energy, April 27–28, 1978. In *Growth and Change*, January 1979.

"Energy analysis and economic valuation." *Southern Economic Journal*, April 1979: 1023–58.

"Methods in economic science." *Journal of Economic Issues* (June 1979): 317–27; (March 1981): 188–93.

"The measure of information: a critique." In *Proceedings of the Third International Congress of Cybernetics and Systems*, ed. J. Rose and C. Bileiu. New York: Springer Verlag, n.d., vol. 3: pp. 187–217.

"Energy, matter, and economic valuation: where do we stand?" In *Energy, Economics and the Environment: Conflicting Views of an Essential Interrelationship*, eds. Herman E. Daly and Alvaro F. Umaña. Boulder, Col.: Westview Press, 1981.

"Energetic dogma, energetic economics, and viable technologies." in *Advances in the Economics of Energy and Resources*, vol. 4, ed. J.R. Maroney. Greenwich, Conn.: JAI Press, 1982.

"Feasible recipes versus viable technologies." *Atlantic Economic Journal* (October 1983): 21–31.

"Hermann Heinrich Gossen: his life and work in historical perspective." In *The Laws of Human Relations and the Rules of Human Actions Derived Therefrom*, ed. R.C. Blitz, Jr. Cambridge, Mass.: MIT Press, 1983, pp. 11–140.

"An epistemological analysis of statistics: the science of collective description and of rational guessing." In *Studies in Probability and Related Topics*, eds. M.M.C. Demetrescu and M. Iosifescu. Montreal: Nagrad, 1983, pp. 221–59.

"Man and production." in *Foundations of Economics*, eds. M. Baranzini and R. Scazzieri. Oxford: Blackwell, 1986, pp. 247–80.

"Entropy." In *The New Palgrave Dictionary*, vol. II, London: Macmillan, 1987.

"Interplay between institutional and material factors." in *Barriers to Full Employment*, eds. J.A. Kregel, Egon Matzner, and Allessandrdo Roncaglia. London: Macmillan Press, 1988, pp. 297–326.

"An emigrant from a developing country: autobiographical notes." *Banca Nazioanle del Lavoro Quarterly Review*, no. 164 (March 1988); no. 184 (March 1993).

"Closing remarks: about economic growth – a variation on a theme by David Hilbert." *Economic Development and Cultural Change*, **36**, suppl. 3 (April 1988): S291–S307.

"Thermodynamics, economics, and information." In *Organization and Change in Complex Systems*, ed. M. Alonso. New York: Paragon House, 1990.

"Nicholas Georgescu-Roegen about himself." In *Eminent Economists: Their Life Philosophies*, ed. M. Szenberg. Cambridge: Press Syndicate of the University of Cambridge, 1992, pp. 128–159.

"Looking back." In *Entropy and Bioeconomics*: Proceedings, eds. J. Dragan, E. Seifert, and M. Demetrescu. Milan: European Association for Bioeconomic Studies, 1993, pp. 11–21.

"Thermodynamics and we, the humans." In *Entropy and Bioeconomics*: Proceedings, eds. J. Dragan, E. Seifert and M. Demetrescu. Milan: European Association for Bioeconomic Studies, 1993, pp. 184–201.

References

Adelman, F. (1972), "The Entropy Law and the economic process" (book review), *Journal of Economic Literature*, **10**: 458–60.

Allen, R.G.D., (1932) "The foundations of a mathematical theory of exchange," *Economica*, **12** May: 197–226.

Artstein, Z. (1980), "Generalized solutions to continuous-time allocation processes," *Econometrica* **48**: 899–922.

Atkins, P.W. (1984), *The Second Law*, New York: Scientific American/W.H. Freeman.

Battalio, R., L. Green, and J. Kagel (1981), "Income–leisure tradeoffs of animal workers," *American Economic Review*, **7**: 621–32.

Battalio, R., J. Kagel, and D. MacDonald (1985), "Animal's choices over uncertain outcomes: some initial experimental results," *American Economic Review*, **75**: 597–613.

Beaud, M. and G. Dostaler (1995), *Economic Thought Since Keynes*, Aldershot, UK: Edward Elgar.

Betancourt, R. (1986), "A generalization of modern production theory", *Applied Economics*, 18: 915–28.

Biswas, A.K., and W.G. Davenport (1980), *Extractive Metallurgy of Copper*, 2nd edn, Oxford: Pergamon Press; 3rd edn, 1994.

Blaug, M. (1985), *Economic Theory in Retrospect*, 4th edn, Cambridge: Cambridge University Press.

Bliss, C.J. (1975), *Capital Theory and the Distribution of Income*, Amsterdam: North-Holland.

Bauman, M. ed. (1958), *Expectations, Uncertainty and Business Behavior*, New York: Social Science Research Council.

Boyer, C. (1985), *A History of Mathematics*, Princeton, N.J.: Princeton University Press.

Braudel, F. (1979), *The Structures of Everyday Life*, vol. 1, New York: Harper & Row.

Bromley, D., ed. (1995), *The Handbook of Environmental Economics*, Oxford: Basil Blackwell.

Burmeister, E. (1980), *Capital Theory and Dynamics*, Cambridge: Cambridge University Press.

Chapman, P.F. and F. Roberts (1983), *Metal Resources and Energy*, London: Butterworth.

Chipman, J., L. Hurwicz, M. Richter, and H. Sonnenschein, eds (1971), *Preferences, Utility, and Demand*, New York: Harcourt.

Costanza, R., ed. (1991), *Ecological Economics*, New York: Columbia University Press.

Cottrell, A.H. (1967), *An Introduction to Metallurgy*, New York: St. Martin's Press.

Coveney, P. and R. Highfield (1990), *The Arrow of Time*, New York: Fawcett.

Daly, H. (1968), "On economics as a life science," *Journal of Political Economy*, **76**: 342–406.

Daly, H. (1992), "Is entropy law relevant to the economics of natural resource scarcity? – Yes, of course it is!" *Journal of Environmental Economics and Management*, **23**: 91–5.

Daly, H. (1993), "The steady-state economy: toward a political economy of bio-physical equilibrium and moral growth," in H. Daly and K. Townsend, eds, *Valuing the Earth: Economics, Ecology, Ethics*. Cambridge, Mass.: MIT Press.

Daly, H. (1995), "On Nicholas Georgesçu-Roegen's contributions to economics: an obituary essay," *Ecological Economics*, **13**: 149–54.

Daly, H. (1997), "Georgesçu-Roegen versus Solow–Stiglitz," *Ecological Economics*, **22**: 261–6.

Daly, H. and J. Cobb (1989), *For the Common Good: Redirecting the Economy toward Community, the Environment, and a Sustainable Future*, Boston, Mass.: Beacon Press.

Daly, H. and K. Townsend, eds (1993), *Valuing the Earth: Economics, Ecology, Ethics*. Cambridge, MA: MIT Press.

Dasgupta, P. and G. Heal (1979), *Economic Theory of Exhaustible Resources*, Cambridge: Cambridge University Press.

Dugdale, J.S. (1996), *Entropy and its Physical Meaning*, Bristol, Penn.: Taylor & Francis.

Ellis, O. (1978), *Copper and Copper Alloys*, Cleveland: American Society for Metals.

Faber, M., H. Niemes, and G. Stephan (1987), *Entropy, Environment, and Resources*, Berlin: Springer; 2nd edn, 1995.

Fermi, Enrico (1937/1956), *Thermodynamics*, New York: Dover Publications.

Frautschi, S. (1982), "Entropy in an expanding universe," *Science*, **217**: 593–99.

Fry, I. (1995), "Evolution in thermodynamic perspective: a historical and philosophical angle," *Zygon*, **30**: 227–48.

Gale, D. (1960), "A note on revealed preference," *Economica*, Nov.: 348–54.

Gaskell, D. (1981), *Introduction to Metallurgical Thermodynamics*, 2nd edn, New York: Hemisphere Publishing Corporation.

Gayer, T. (1997), "Hazardous waste risks, housing prices, and economic methodology", PhD dissertation, Duke University.

Goldstein, M. and I. Goldstein (1993), *The Refrigerator and the Universe: Understanding the Laws of Energy*, Cambridge, Mass.: Harvard University Press.

Guillen, M. (1995), *Five Equations that Changed the World*, New York: Hyperion.

Hawkins, D. and H. Simon (1949), "Note: some conditions of macroeconomic stability," *Econometrica*, **17**: 245–8.

Hey, J. and P. Lambert (1987), *Surveys in the Economics of Uncertainty*, Oxford: Basil Blackwell.

Hotelling, H. (1931), "The economics of exhaustible resources," *Journal of Political Economy*, **39**: 137–75.

Houthakker, H. (1950), "Revealed preference and the utility function," *Econometrica*, May: 159–74.

Howarth, R. and R. Norgaard (1995), "Intergenerational choices under global environmental change," in D. Bromley, ed., *Handbook of Environmental Economics*, Cambridge, Mass.: Basil Blackwell.

Hubbard, L. (1939), *The Economics of Soviet Agriculture*, London.

Kaplansky, I. (1977), *Set Theory and Metric Spaces*, New York: Chelsea Publishing.

Keenan, C., J. Wood, and D. Kleinfelter (1976), *General College Chemistry*, 5th edn, New York: Harper & Row.

King, E., A. Mah, and L.B. Pankratz (1973), *Thermodynamic Properties of Copper and its Inorganic Compounds*. Unknown place of publication: International Copper Research Association (INCRA).

Klein, C. (1980), "Modeling rate of production and time utilization of plant: a flow-fund duality approach," PhD dissertation, University of North Carolina, Chapel Hill.

Knight, F. (1921), *Risk, Uncertainty, and Profit*, Boston: Houghton Mifflin.

Koopmans, T.C., ed. (1951), *Activity Analysis of Production and Allocation*, New York: John Wiley.

Kotas, T.J. (1985), *The Exergy Method of Thermal Plant Analysis*, London: Butterworth.

Leontief, W. (1941), *The Structure of the American Economy*, Cambridge, Mass.: Harvard University Press.

Lodge, O. (1929), *Energy.* New York: Robert M. MacBride.

Lozada, G. (1995), "Georgesçu-Roegen's defense of classical thermodynamics revisited," *Ecological Economics*, **12**: 31–44.

Lozada, G. (1999), "The role of entropy and energy in natural resource economics," in J. Goudy and K. Mayumi, eds, *Bioeconomics and Sustain-*

ability: Essays in Honor of Nicholas Georgesçu-Roegen, Cheltenham, UK: Edward Elgar, forthcoming.

Mackowiak, J. (1965), *Physical Chemistry for Metallurgists*, New York: American Elsevier.

Marshall, A. (1997), *Principles of Economics*, Amherst, NY: Prometheus Books; first published 1890.

Mayumi, K. (1992), "The new paradigm of Georgesçu-Roegen and the tremendous speed of increase in entropy in the modern economic process," *Journal of Interdisciplinary Economics*, **4**: 101–17.

Mayumi, K. (1993), "Georgesçu-Roegen's 'Fourth Law of Thermodynamics', the modern energetic dogma, and ecological salvation", in L. Bonatti et al. eds Trends in Physical Chemistry, New York: Elsevier, pp. 351–64.

Mayumi, K. (1995), "Nicholas Georgesçu-Roegen (1906–1994): an admirable epistemologist," *Structural Change and Economic Dynamics*, **6**: 261–5.

Mayumi, K. (1997), "Information, pseudo measures and entropy: an elaboration on Nicholas Georgesçu-Roegen's critique," *Ecological Economics*, **22**: 249–59.

McTaggart, J. (1927), *The Nature of Existence*, Cambridge: Cambridge University Press.

Miller, G. (1976), *Chemistry: Principles and Applications*, Belmont, Cal.: Wadsworth.

Mirowski, P. (1989), *More Heat than Light*, Cambridge: Cambridge University Press.

Morowitz, Harold J. (1991), *The Thermodynamics of Pizza: Essays on Science and Everyday Life*, New Brunswick: Rutgers University Press.

Neumann, J. von (1945), "A model of general economic equilibrium," *Review of Economic Studies*, **13**: 1–9.

Pankratz, L.B. (1982), *Thermodynamic Properties of Elements and Oxides*, Washington, D.C.: Bureau of Mines/US Government Printing Office.

Pankratz, L.B., A.D. Mah, and S.W. Watson (1987), *Thermodynamic Properties of Sulfides*, Washington, D.C.: Bureau of Mines/US Government Printing Office.

Papi, U. and C. Nunns, eds, (1969), *Economic Problems of Agriculture in Industrial Societies*, London: Macmillan.

Planck, M. (1927/1991), *The Theory of Heat Radiation*, Mineola, NY: Dover Publications.

Planck, M. (1923), Varlesungen veber die Theorie der Waermestrahlung. Leipzig: Johann Barth.

Prigogine, I. and I. Stengers (1984), *Order Out of Chaos*, New York: Bantam Books.

Rao, Y.K. (1985), *Stoichiometry and Thermodynamics of Metallurgical Processes*, Cambridge: Cambridge University Press.

Rees, R. (1987), "The theory of principal and agent: Parts I and II," in J. Hey and P. Lambert, eds, *Surveys in the Economics of Uncertainty*, Oxford: Basil Blackwell.

Robinson, A. (1974), Nonstandard Analysis. Amsterdam: North Holland.

Russell, B. (1903), *Principles of Mathematics*, Cambridge: Cambridge University Press.

Samuelson, P. (1983), *Foundations of Economic Analysis*, enlarged edn, Cambridge, Mass.: Harvard University Press.

Samuelson, P. (1950a), "The problem of integrability in utility theory," *Economica*, N.S. **17**: 355–85.

Samuelson, P. (1950b), "Probability and attempts to measure utility," *Economic Review* (Hitosubashi University) I: 167–73.

Schulman, L. (1997), *Time's Arrows and Quantum Measurement*, Cambridge: Cambridge University Press.

Schumpeter, J. (1954), *History of Economic Analysis*, New York: Oxford University Press.

Schumpeter, J. (1964), *Business Cycles*, abridged edn, New York: McGraw-Hill; first published 1939.

Screpanti, E. and S. Zamagni (1993), *An Outline of the History of Economic Thought*, Oxford: Clarendon Press.

Smallman, R.E. (1976), *Modern Physical Metallurgy,* 3rd edn, London: Butterworth.

Smith, V.K., ed. (1979), *Scarcity and Growth Reconsidered*, Baltimore, Md: Johns Hopkins University Press.

Solow, R. (1974), "Intergenerational equity and exhaustible resources," *Review of Economic Studies Symposium*, **41/42**: 29–45.

Solow, R. (1974), "The economics of resources and the resources of economics." *American Economic Review*, **64** (2) May: 1–14.

Stiglitz, J. (1970), "Reply to Mrs Robinson on the choice of technique," *Economic Journal*, **80**: 420–2.

Stiglitz, J., "A Neoclassical Analysis of the Economics of Natural Resources", in ed. V. Smith, Scarcity and Growth Reconsidered, Baltimore, Md: Johns Hopkins University Press.

Sweezy, P. (1942), *The Theory of Capitalist Development*, New York.

Szenberg, M. (1992), *Eminent Economists*, Cambridge: Cambridge University Press.

Tang, A., F. Westfield, and J. Worley (1976), *Evolution, Welfare, and Time in Economics: Essays in Honor of Nicholas Georgesçu-Roegen*, Lexington, Mass.: Lexington Books.

Toman, M., J. Pezzey, and J. Krautkraemer (1995), "Neoclassical economic growth theory and sustainability," ed. D. Bromley, in *The Handbook of Environmental Economics*, Oxford: Basil Blackwell, pp. 139–65.

Varian, H. (1992), *Microeconomic Analysis*, 3rd edn, New York: W.W. Norton.

Varian, H. (1996), *Intermediate Microeconomics: A Modern Approach*, 4th edn, New York: W.W. Norton.

Weintraub, R., S. Mearden, T. Gayer, and H. Banzhaf (1998), "Archiving the history of economics," *Journal of Economic Literature*, September: 1496–1512.

Wicken, J. (1987), *Evolution, Thermodynamics, and Information: Extending the Darwinian Program*, New York: Oxford University Press.

Young, J. (1991), "Is the Entropy Law relevant to the economics of natural resource scarcity?" *Journal of Environmental Economics and Management*, **21**: 169–79.

Zemansky, Mark W. (1968), *Heat and Thermodynamics: An Intermediate Textbook*, 5th edn, New York: McGraw-Hill.

Index